SETS AND NUMBERS

BY

S. ŚWIERCZKOWSKI

Senior Research Fellow,
Institute of Advanced Studies
Australian National University

ROUTLEDGE & KEGAN PAUL
LONDON AND BOSTON

First published 1972
by Routledge & Kegan Paul Ltd
Broadway House, 68–74 Carter Lane
London EC4V 5EL and
9 Park Street,
Boston, Mass. 02108, U.S.A.

Printed in Hungary

ISBN 0 7100 7137 x

Contents

Preface

1. Sets

1 Sets and elements 1
2 How to denote sets 1
3 Logical notation 4
4 Inclusion, subset 4
5 Intersection 5
6 The one-element set and the empty set 7
7 Union 8
8 Difference 10
Exercises 11

2. Products and Mappings

1 Ordered pairs and triplets 13
2 Product 14
3 Mapping (function) 15
4 Surjection, injection and bijection 18
5 Inverse mapping 20
6 Composite mapping 22
Exercises 23

3. Cardinal Numbers

1 The postulate of cardinal numbers 25
2 The cardinal numbers $\aleph_0$ and c 27

3 Inequality 32
4 Addition 35
5 Infinite sets 39
6 The Cantor–Bernstein–Schröder theorem 42
Exercises 46

4. Cardinal Algebra

1 Product 48
2 Exponentiation 52
3 The set of all subsets of A 56
4 Sequences 59
Exercises 61

About Wrong Definitions 63

Hints and Answers 65

Index 74

Preface

The aim of this book is to create familiarity with the most basic operations on sets and functions. I wrote it in the years 1963–5 while teaching the subject at the University of Sussex. The delay in the publication arose from a mistake in the postal delivery of the manuscript.

I am grateful to Professor W. Ledermann for his many comments which have led to an improvement of this presentation.

Canberra S. Świerczkowski

CHAPTER ONE

Sets

1. Sets and elements

Georg Cantor (1845–1918) was the first mathematician to study sets systematically. The intuitive meaning of 'a set' is the same as that of a collection, a class or a family. With a set one usually associates certain objects which are called elements or members of that set. We shall in this book assume the intuitive meaning of 'set' and 'element'.

It is customary to write the symbol $\in$ instead of 'is an element of' or 'belongs to'. Thus $x \in A$ means 'x is an element of A' or 'x belongs to A'. If this notation is used, it is of course assumed that A is a set.

2. How to denote sets

Certain sets are finite; they have finitely many elements. Such a set can be denoted simply by listing all its elements, for example $\{a, b, c\}$ denotes the set whose elements are the three letters a, b and c. The same set can be denoted by $\{b, c, a\}$, $\{c, a, b\}$ and in three other ways, because all that we wish to write between the brackets are the elements of the set (or, more precisely, their names) and it does not matter in which order we list them.

This notation is sometimes generalized to infinite sets. We

shall call 0, 1, 2, 3, ... the *natural numbers* and we shall denote the set of all natural numbers by N. Let E be the set of natural numbers divisible by 2. Then we can write

$$N = \{0, 1, 2, \ldots\}, \qquad E = \{0, 2, 4, \ldots\}.$$

Similarly the set of all squares can be denoted by $\{0, 1, 4, \ldots\}$. But this notation 'with three dots' is always ambiguous, it is only a guess that 16 belongs to $\{0, 1, 4, \ldots\}$.

A more precise notation can be used in the following situation: Suppose we have a set A and a property p which is meaningful for every element x of A. Such a property is a statement about x and therefore we shall denote it by $p(x)$ rather than by p. By saying that $p(x)$ is *meaningful* for every x in A we mean that the statement $p(x)$ is either true or false, depending on the choice of x, for every element x. Then we shall denote by

$$\{x \in A \mid p(x)\}$$

the set of all those x belonging to A for which $p(x)$ is true. Then

$$E = \{x \in N \mid \text{there exists a } y \in N \text{ such that } x = 2y\},$$

or, in words, E is the set of all natural numbers x which have the property 'there exists a y belonging to N such that $x = 2y$', and this notation is certainly more precise than $\{0, 2, 4, \ldots\}$.

We shall denote by R the set of all real numbers. Then R includes all the integers $0, \pm 1, \pm 2, \ldots$, the rational numbers like $3/2$, $-5/7$, and irrational numbers like $\sqrt{2}, \sqrt{3}, \pi, e$. A convenient way of thinking about R is the following: We imagine a straight line (Figure 1) with two points marked on it, one of them denoted by 0, the other by 1. Then every integer (whole number) has its place on this line, for example 2 is the point whose distance from 0 is twice the distance from 1 to 0, -4 lies on the opposite side of 0 from 1 at a distance four times larger, and so on. Also every rational number has its place on the line,

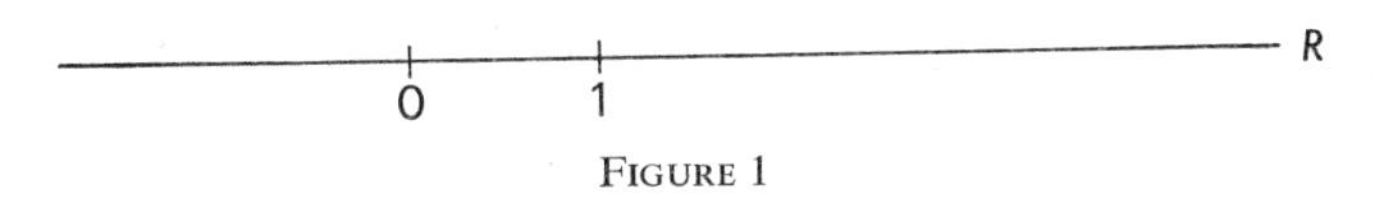

FIGURE 1

for we can find easily 1/2, 1/3, 1/4,..., and the other rational numbers are (positive and negative) multiples of these. Finally, there are points on this line which do not correspond to any rational number, such as, for example, the point to the right of 1 whose distance from 0 is equal to the length of the diagonal of a square with sides of length 1; this point is called $\sqrt{2}$. In general, we shall call every point on this line a *real number*, and these will be all the real numbers. The set R will also be called *the line* or *the real line*.

If $a, b \in R$, then the set of all points lying between a and b is called an *interval*. We shall use two kinds of interval, namely

$$\{x \in R \mid a < x < b\} \quad \text{and} \quad \{x \in R \mid a \leqslant x \leqslant b\}$$

which are called the *open* and the *closed* intervals between a and b; they are also denoted by (a, b) and $[a, b]$ respectively.

We shall say that two sets A and B are *identical* and write $A = B$ if and only if A and B have the same elements. In other words $A = B$ if every element of A belongs to B and vice versa. For example

$$[-1, 1] = \{y \in R \mid \text{there exists an } x \in R \text{ such that } y = \sin x\}.$$

3. Logical notation

We shall use certain standard abbreviations which should be read as follows:

$\exists$ there exists
$\forall$ for every
$\Rightarrow$ implies
$\Leftrightarrow$ if and only if (or, is equivalent to)
s.t. such that
$\in$ is an element of

If α and β are any statements then, according to the above, $\alpha \Leftrightarrow \beta$ should be read as 'α is equivalent to β', or 'α if and only if β'. This means precisely the same as saying that 'α implies β and β implies α'.

4. Inclusion, subset

A part of a set is called a subset of that set, or more precisely

DEFINITION 1.1. Let A, B be sets. We say that A is a *subset* of B, denoting this by $A \subset B$, if

$$\forall x\colon x \in A \Rightarrow x \in B.$$

In words this says that every object x has the property that if it belongs to A then it must also belong to B (see Figure 2). We also say that A is *contained in B* and we call $\subset$ the *inclusion* sign*. If E, N, R denote the sets as considered in Section 2,

* The sign $\subseteq$ is sometimes used instead of $\subset$ (see J. A. Green, *Sets and Groups*, Routledge & Kegan Paul, 1965).

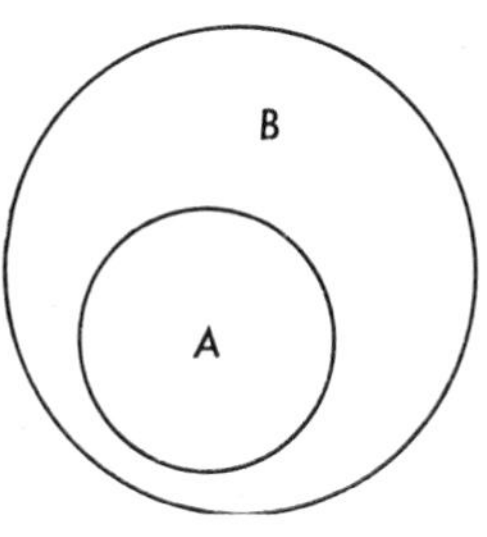

FIGURE 2

then we have

$$\{2, 4, 6\} \subset E, \qquad E \subset N \qquad \text{and} \qquad N \subset R.$$

It follows from Definition 1.1 that for every set A we have $A \subset A$.

THEOREM 1.1. *$A \subset B$ and $B \subset A \Leftrightarrow A = B$.*

Proof. To say $A = B$ means the same as to say that every element of A is also an element of B and every element of B is also an element of A, that is $A \subset B$ and $B \subset A$.

5. Intersection

The common part of two sets A and B will be denoted by $A \cap B$ and called their *intersection*. This is defined as follows

DEFINITION 1.2. Let A and B be sets. Then $A \cap B$ is a set such that

$$\forall x: x \in A \cap B \Leftrightarrow x \in A \quad \text{and} \quad x \in B.$$

That is, x is an element of $A \cap B$ if and only if x is an element of both A and B.

EXAMPLES

$$\{a, b, c\} \cap \{e, f, b, d, c\} = \{b, c\},$$
$$E \cap \{x \in N \mid \exists y \in N \text{ s.t. } x = 3y\} = \{0, 6, 12, \ldots\},$$
$$\{x \in R \mid 0 < x \leqslant 1\} \cap \{x \in R \mid \tfrac{1}{2} \leqslant x < 2\} = [\tfrac{1}{2}, 1],$$

and in Figure 3, $A \cap B$ is the shaded area.

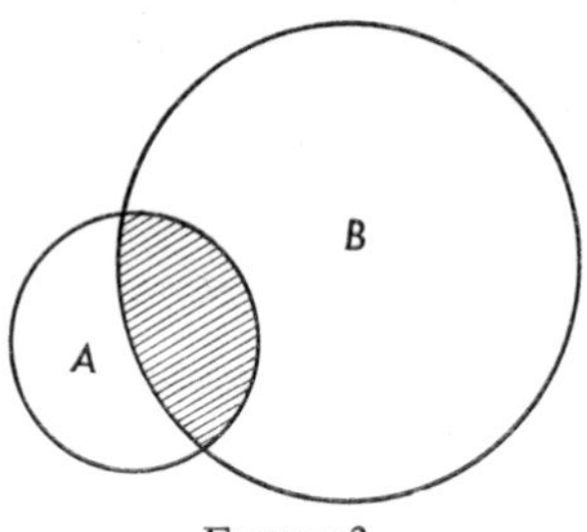

FIGURE 3

Now let $A_0, A_1, A_2, \ldots, A_n$ be $n+1$ sets. The set of all those elements which belong to each of the sets $A_0, \ldots, A_n$ is denoted by $\bigcap_{k=0}^{n} A_k$. The definition of this intersection is

DEFINITION 1.3.

$$\forall x: x \in \bigcap_{k=0}^{n} A_k \Leftrightarrow x \in A_k, \quad \text{for every } k \text{ in} \quad \{0, 1, 2, \ldots, n\}.$$

Now suppose that we have associated with every natural number a set. Let, for every natural k, A_k denote the set associated with k. Then we say that we have defined a sequence of sets $A_0, A_1, A_2, \ldots$. The intersection of all sets A_k is denoted by $\bigcap_{k=0}^{\infty} A_k$ and is defined by

DEFINITION 1.4.

$$\forall x \colon x \in \bigcap_{k=0}^{\infty} A_k \Leftrightarrow x \in A_k, \quad \text{for every } k \text{ in} \quad \{0, 1, 2, 3, \ldots\}.$$

EXAMPLE. If $A_k = \left(-\frac{1}{k+1}, 1+\frac{1}{k+1}\right)$, then $\bigcap_{k=0}^{\infty} A_k = [0, 1]$.

6. The one-element set and the empty set

The sets $\{a, b\}$ and $\{b, c\}$ have exactly one element in common, that is, the set $\{a, b\} \cap \{b, c\}$ has one element b. According to our convention in Section 2 we shall denote this set by $\{b\}$, so that

$$\{a, b\} \cap \{b, c\} = \{b\}.$$

Now consider the set $\{1, 2\} \cap \{3, 4\}$. Is it a set? It has no elements, but on the other hand we have already decided to assign a set $A \cap B$ to any sets A and B, and therefore we must call $\{1, 2\} \cap \{3, 4\}$ a set. We shall call it the *empty set*. This empty set is denoted by $\varnothing$ and, though it was introduced here in a special way, it is properly defined as follows:

DEFINITION 1.5. $\varnothing$ is a set such that $x \in \varnothing$ is false for every x (in other words, a set without elements).

We can now say that $\{1, 2\} \cap \{3, 4\} = \varnothing$. In general two sets A, B will be called disjoint if they have no elements in common:

DEFINITION 1.6. A and B are *disjoint* $\Leftrightarrow A \cap B = \varnothing$.

Let us observe that if $a, b \in R$, then the interval

$$[a, b] = \{x \in R \mid a \leqslant x \leqslant b\}$$

is non-empty if and only if $a \leqslant b$, otherwise $[a, b] = \varnothing$.

7. Union

The set which consists of all elements of A and all elements of B is called the *union* of A and B, and is denoted by $A \cup B$. Formally

DEFINITION 1.7.

$$\forall x \colon x \in A \cup B \Leftrightarrow x \in A \quad \text{or} \quad x \in B.$$

EXAMPLES

$$\{a, b, c\} \cup \{e, b, f, d, c\} = \{a, b, c, e, f, d\},$$

$$\{x \in R \mid 0 < x \leqslant 1\} \cup \{x \in R \mid \tfrac{1}{2} \leqslant x < 2\} = (0, 2),$$

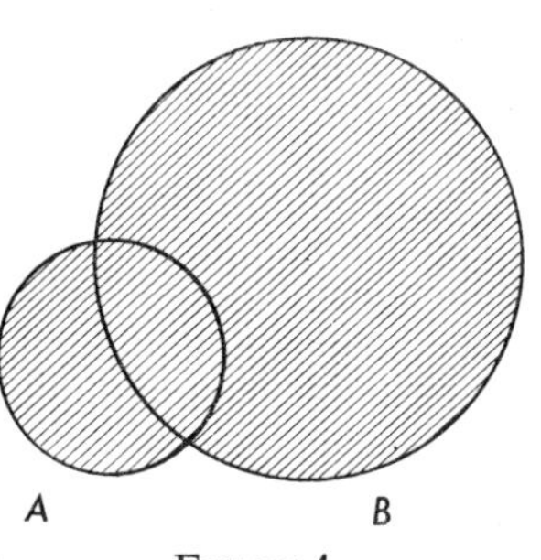

FIGURE 4

and in Figure 4 the union of the two discs A and B is the shaded area.

If $A_0, A_1, A_2, \ldots, A_n$ are $n+1$ sets, the set of all those elements which belong to at least one of the sets $A_0, \ldots, A_n$ is denoted by $\bigcup_{k=0}^{n} A_k$ and it is called the union of the sets $A_0, \ldots, A_n$. In other words

DEFINITION 1.8.

$$\forall x: x \in \bigcup_{k=0}^{n} A_k \Leftrightarrow \exists k \in \{0, 1, 2, \ldots, n\} \text{ s.t. } x \in A_k.$$

The union of a sequence of sets $A_0, A_1, A_2, \ldots,$ is defined similarly by

DEFINITION 1.9.

$$\forall x: x \in \bigcup_{k=0}^{\infty} A_k \Leftrightarrow \exists k \in N \text{ s.t. } x \in A_k.$$

EXAMPLE

If $A_k = \left[\frac{1}{k+1}, 1-\frac{1}{k+1}\right]$, then $\bigcup_{k=0}^{\infty} A_k = (0, 1)$.

8. Difference

If A and B are two sets then the set of all those elements of A which do not belong to B is denoted by $A-B$. We shall write $\notin$ to denote 'does not belong to'. Then the *difference* is defined by

DEFINITION 1.10.

$$A-B = \{x \in A \mid x \notin B\}.$$

In Figure 5, $A-B$ is the shaded area.

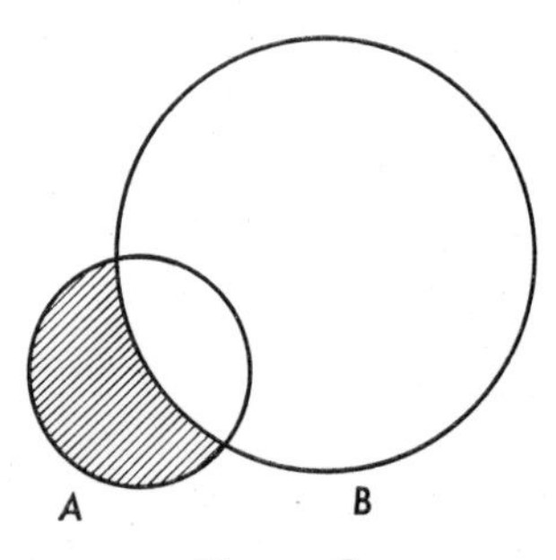

FIGURE 5

THEOREM 1.2. $A - B = \varnothing \Leftrightarrow A \subset B$.

Proof. $A - B = \varnothing$ means that there is no x which belongs to A and not to B, that is for every x, we have $x \in A \Rightarrow x \in B$. This means that $A \subset B$.

EXERCISES

1. Obtain eight distinct sets by starting from the sets $\{1, 2\}$ and $\{2, 3\}$ and then using the operations $\cap$, $-$, $\cup$.

2. Show that the set N of natural numbers can be written as a union $N = \bigcup_{k=0}^{\infty} A_k$ of infinitely many sets $A_0, A_1, A_2, \ldots$ such that any two of these sets are disjoint (that is, $A_k \cap A_l = \varnothing$ if $k \neq l$) and each of them is infinite.

3. For any finite set A, denote by $|A|$ the number of elements of A. If $|A| = 9$, $|B| = 7$, $|A \cap B| = 3$, find $|A \cup B|$.

4. Find a formula valid for any finite sets A, B which gives $|A \cup B|$ in terms of $|A|$, $|B|$ and $|A \cap B|$.

5. Find a similar formula as in Exercise 4 which gives $|A \cup B \cup C|$ in terms of $|A|$, $|B|$, $|C|$, $|A \cap B|$, $|B \cap C|$, $|A \cap C|$, $|A \cap B \cap C|$.

6. In a battle, 70% of the combatants lost one eye, 80% an ear, 75% an arm, 85% a leg, x% lost all 4. Find the minimum value that is possible for x.

7. In a survey of 100 students the numbers studying the various languages were found to be: Spanish 28, German 30, French 42, Spanish and German 8, Spanish and French 10, German and French 5, all three languages 3.

(i) How many were studying no language?
(ii) How many had French as their only language?

8. Prove that

(i) $A \cap (B \cup C) = (A \cap B) \cup (A \cap C)$,

(ii) $A \cup (B \cap C) = (A \cup B) \cap (A \cup C)$.

9. Indicate which of the statements

$$A \cap B = A, \qquad A \cup B = B, \qquad A \cup B = A \cap B,$$
$$A - B = \varnothing, \qquad (A-B) \cup (B-A) = \varnothing$$

are equivalent to $A \subset B$ and which are equivalent to $A = B$.

10. Give an example of a sequence of open intervals A_0, A_1, $A_2, \ldots$ such that any finite number of these has a non-empty intersection (that is, $\bigcap_{k=0}^{n} A_k \neq \varnothing$ for every $n \in \{0, 1, 2, \ldots\}$), but the intersection of all these intervals is empty (that is, $\bigcap_{k=0}^{\infty} A_k = \varnothing$).

Remark. It is impossible to find a sequence of closed intervals with the above property.

CHAPTER TWO

Products and Mappings

1. Ordered pairs and triplets

Suppose that a, b are any two objects. We shall assume that there is an object denoted by $\langle a, b\rangle$ which we shall call the *ordered pair* with *first term* a and *second term* b. Two ordered pairs $\langle a_1, b_1\rangle$ and $\langle a_2, b_2\rangle$ will be called equal if and only if their first terms coincide and their second terms coincide, that is

$$\langle a_1, b_1\rangle = \langle a_2, b_2\rangle \Leftrightarrow a_1 = a_2 \quad \text{and} \quad b_1 = b_2.$$

Note that the ordered pair $\langle a, b\rangle$ is not the same thing as the set $\{a, b\}$ because $\langle a, b\rangle$ and $\langle b, a\rangle$ are two distinct ordered pairs, whereas $\{a, b\}$ and $\{b, a\}$ denote the same set. Also, we allow the ordered pair $\langle a, a\rangle$, but $\{a, a\}$ is just another way of writing $\{a\}$.

Similarly, we shall postulate that for any objects a, b, c there exists an object denoted by $\langle a, b, c\rangle$ which we shall call the *ordered triplet* with *first term* a, *second term* b and *third term* c. Two ordered triplets $\langle a_1, b_1, c_1\rangle$ and $\langle a_2, b_2, c_2\rangle$ will be called equal if and only if $a_1 = a_2$ and $b_1 = b_2$ and $c_1 = c_2$. Thus $\langle a, b, c\rangle \neq \langle a, c, b\rangle$. Note that an ordered triplet of the form $\langle a, b, b\rangle$ is different from the ordered pair $\langle a, b\rangle$ (because there is no third term in $\langle a, b\rangle$, and we have postulated that there is always a third term in every ordered triplet).

2. Product

If A and B are any sets, then the set of all ordered pairs $\langle x, y\rangle$ such that $x \in A$ and $y \in B$ is denoted by $A \times B$ and it is called the *product of A and B*.

EXAMPLE

$$\{1, 2\} \times \{a, b, c\} = \{\langle 1, a\rangle, \langle 2, a\rangle, \langle 1, b\rangle, \langle 2, b\rangle, \langle 1, c\rangle, \langle 2, c\rangle\}.$$

The set $R \times R$ is the set of all pairs $\langle x, y\rangle$ where x and y are any real numbers. In coordinate geometry every such pair $\langle x, y\rangle$ represents a point of the Euclidean plane. Therefore we shall call $R \times R$ simply *the plane*. The subsets of $R \times R$

$$\{\langle x, y\rangle \in R \times R \mid y = 0\}, \qquad \{\langle x, y\rangle \in R \times R \mid x = 0\}$$

are called the x-axis and the y-axis. Every element $\langle x, y\rangle \in R \times R$ is called a *point* of the plane.

Let A be a finite set. Recall the notation (introduced in Exercise 3, Chapter 1)

$$|A| = \text{the number of elements of } A.$$

THEOREM 2.1. *If A, B are finite, then* $|A \times B| = |A| . |B|$.

Proof. For every x in A there are $|B|$ choices of y in B such that $\langle x, y\rangle \in A \times B$. As there are $|A|$ choices for x in A, we finally get $|A| . |B|$ pairs in $A \times B$.

3. Mapping (function)

In the later stages of this book we shall be able to compare two infinite sets and say whether they have the same number of elements or that one has more elements than the other. The basic idea used in such comparisons is that of a mapping. The words 'mapping' and 'function' are often used interchangeably in mathematics and we shall now define more precisely what is meant by the word 'mapping' in this book.

Let A and B be any two sets. Intuitively we say that we have a mapping (or function) from A to B (written $f{:}A \to B$) if we have a 'law' assigning to every element x in A a corresponding element y in B. (We call y the *image* of x under this mapping). Thus a mapping is composed of three objects: two sets A, B and a certain 'law' which can be symbolized by f. But since we wish to regard a mapping as a single object, it is most natural to identify it with the ordered triplet $\langle f, A, B\rangle$. The 'law' f can be defined precisely in terms of the ideas of subset and product of sets, as follows. Suppose that we have a subset of $A \times B$ and we denote this subset by f (we have hitherto denoted sets by capital letters but this is not a rule). Suppose further that f has the following property

$$\forall x \text{ in } A, \exists \text{ exactly one } y \text{ in } B \text{ s.t. } \langle x, y\rangle \in f. \qquad (1)$$

Then f gives us the following law: assign to every x in A the unique y in B such that $\langle x, y\rangle \in f$. We define now the notion of a mapping as follows

DEFINITION 2.1. If A, B are sets and f is a subset of $A \times B$ such that f has the above property (1), then the ordered triplet

$\langle f, A, B\rangle$ is called a *mapping* from A to B. The mapping $\langle f, A, B\rangle$ is also denoted by

$$f: A \longrightarrow B \quad \text{or} \quad A \longrightarrow B.$$

We read $f: A \to B$ (or $A \xrightarrow{f} B$) as *f maps A into B*.

EXAMPLE

If
$$A = \{1, 2, 3\}, \quad B = \{a, b\} \quad \text{and}$$
$$f = \{\langle 1, a\rangle, \quad \langle 2, a\rangle, \quad \langle 3, b\rangle\}$$

then $f: A \to B$ (verify condition (1)).

DEFINITION 2.2. Let $f: A \to B$ be a mapping. If a pair $\langle x, y\rangle$ in $A \times B$ is such that $\langle x, y\rangle \in f$, then we say that y is the image of x, and we denote the element y also by $f(x)$.*

In the above example $\langle 1, a\rangle \in f$, thus a is the image of 1 and we can write $a = f(1)$. Similarly $a = f(2)$, $b = f(3)$.

Let A be a disc, let B be a line (Figure 6) which does not intersect the disc and let f be the subset of $A \times B$ consisting of all those pairs $\langle x, y\rangle$, where $x \in A$, $y \in B$ and the line through x and y is perpendicular to B. Then $f: A \to B$ (check condition (1)). We may call this mapping a projection.

The functions of a real variable, as discussed in calculus, are mappings—for example, if

$$f = \{\langle x, y\rangle \in R \times R \mid x^2 = y\},$$

then $f: R \to R$ (check (1)!); we have $f(x) = x^2$.

* In modern algebra it is also customary to write xf instead of $f(x)$ (see J. A. Green, *op. cit.*).

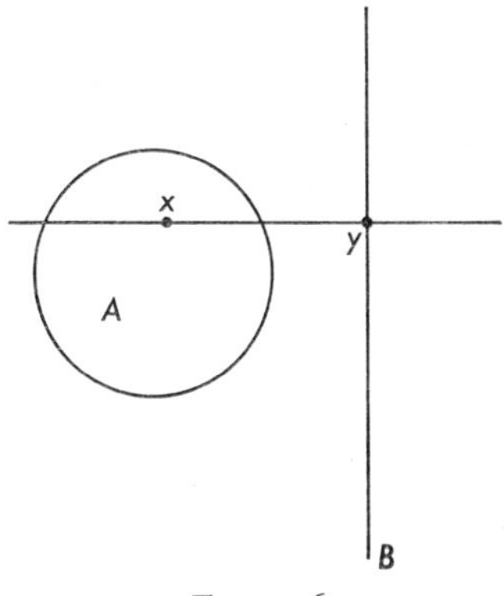

FIGURE 6

For an arbitrary set A denote

$$\iota_A = \{\langle x, y\rangle \in A \times A \mid x = y\}.$$

Then $\iota_A : A \to A$ (we check condition (1) with f and B replaced by ι_A and A). Clearly $\iota_A(x) = x$ holds for every x in A. We call $\iota_A : A \to A$ the *identity mapping* on A.

If we have a mapping $f: A \to B$, we call the set A the *domain* and the set B the *range* of the mapping. We call the subset f of $A \times B$ the *graph* of this mapping. For example, if $f: R \to R$ is as in the above example so that $f(x) = x^2$, then the graph f is a subset of the plane $R \times R$; it is a parabola.

THEOREM 2.2. *If A, B are finite sets then the number of mappings from A to B is $|B|^{|A|}$.*

Proof. Suppose that A has n elements, we denote these by symbols $a_1, a_2, \ldots, a_n$. To define a mapping $f: A \to B$ we have to define $f(a_1), f(a_2), \ldots, f(a_n)$ which should be elements of B. So we have $|B|$ choices for $f(a_1)$, $|B|$ choices for $f(a_2), \ldots, |B|$ choices for $f(a_n)$. Altogether we have $|B|^{|A|}$ choices for $f(a_1), \ldots, f(a_n)$.

4. Surjection, injection and bijection

If we have $f: A \to B$, then we define

$$f(A) = \{f(x) \in B \mid x \in A\}$$
$$= \{y \in B \mid \exists x \in A \text{ s.t. } y = f(x)\}.$$

In other words, $f(A)$ is the set of all images. We always have $f(A) \subset B$.

DEFINITION 2.3. We shall say that $f: A \to B$ is a *surjection* (or, is surjective) if and only if $f(A) = B$.

In other words, $f: A \to B$ is a surjection if and only if every element y of B is the image of some element x of A, that is, $y = f(x)$ for some x. The projection of the disc A in the line B considered above (p. 16) is not a surjection. The two mappings

$$f: R \to R \text{ where } f(x) = x^3,$$
$$g: R \to [-1, 1] \text{ where } g(x) = \sin x,$$

are examples of surjections. Indeed, we have $f(R) = R$ and $g(R) = [-1, 1]$.

There is one essential difference between the graphs f and g in the above examples. We have

$$\langle 0, 0\rangle \in g \text{ and } \langle 2\pi, 0\rangle \in g.$$

Hence there exist distinct numbers $x_1 = 0$ and $x_2 = 2\pi$ which have the same image $g(x_1) = g(x_2) = 0$. This is never the case with f because

$$x_1 \neq x_2 \text{ and } \langle x_1, y_1 \rangle, \langle x_2, y_2 \rangle \in f \Rightarrow y_1 \neq y_2,$$

or equivalently,

$$x_1 \neq x_2 \Rightarrow f(x_1) \neq f(x_2).$$

To distinguish between these two kinds of graphs we introduce the following definition.

DEFINITION 2.4. The graph f of a mapping $f: A \to B$ is called *injective* if different elements of A have different images in B, or in other words,

$$\forall x_1, x_2 \text{ in } A: x_1 \neq x_2 \Rightarrow f(x_1) \neq f(x_2),$$

or equivalently,

$$x_1 \neq x_2 \text{ and } \langle x_1, y_1 \rangle, \langle x_2, y_2 \rangle \in f \Rightarrow y_1 \neq y_2.$$

A mapping whose graph is injective is called an injective mapping (or, an *injection*).

For example the mappings $f: [0, 1] \to [0, 1]$ and $f: [0, 1] \to R$ where

$$f = \{\langle x, y \rangle \in [0, 1] \times [0, 1] \mid x^2 = y\}$$

both have the same graph f which is injective. Hence both mappings are injective.

Of fundamental importance is the situation where $f: A \to B$ is simultaneously an injection and a surjection. Mappings of this kind will be used to compare the 'size' of one infinite set with that of another. With their aid we shall prove that some infinite sets are 'more infinite' than others. It will be convenient to call a mapping that is both an injection and a surjection by one name, a bijection.

DEFINITION 2.5. $f: A \to B$ is a *bijection* (or, is bijective) if $f: A \to B$ is simultaneously a surjection and an injection.

A bijective mapping $f: A \to A$ is called a *permutation* of A. Thus, if A is finite, then there are exactly $|A|!$ bijections $f: A \to A$.

5. Inverse mapping

The importance of bijections lies in the fact that with a bijection $f: A \to B$, we can always associate another bijection $g: B \to A$ such that y is the image of x by one of these mappings if and only if x is the image of y by the other mapping. This mapping $g: B \to A$ is usually denoted by $f^{-1}: B \to A$ and is called the *inverse* of $f: A \to B$. Its existence follows from

THEOREM 2.3. *If $f: A \to B$ is a bijection, then there exists a mapping $f^{-1}: B \to A$ such that*

$$\forall x \in B,\ y \in A: f^{-1}(x) = y \Leftrightarrow f(y) = x,$$

moreover $f^{-1}: B \to A$ is a bijection.

Proof. Let x be an arbitrary element of B. Since $f: A \to B$ is surjective, there exists a y in A such that $x = f(y)$. Moreover, there is only one such y, for if there were y' such that $x = f(y')$ and $y \neq y'$, then we would have $y \neq y'$ and $f(y) = f(y')$ $(= x)$, contrary to the assumption that f is injective. So we can assign to every x in B a unique corresponding y in A such that $x = f(y)$. In this way we have defined a function $f^{-1}: B \rightarrow A$, moreover this function is such that

$$f^{-1}(x) = y \Leftrightarrow f(y) = x.$$

Let us show that $f^{-1}: B \rightarrow A$ is surjective. Let y be an arbitrary element of A. If we denote $f(y)$ by x, then by the above equivalence we have $y = f^{-1}(x)$. This shows that y is an image.

To prove that f^{-1} is injective, we take $x_1 \neq x_2$ in B and we denote $f^{-1}(x_1), f^{-1}(x_2)$ by y_1 and y_2 respectively. By the above equivalence we have $x_1 = f(y_1)$ and $x_2 = f(y_2)$, and thus, as $x_1 \neq x_2$, we must have $y_1 \neq y_2$, and so $f^{-1}(x_1) \neq f^{-1}(x_2)$.

Later on we shall use mainly the following consequence of Theorem 2.3.

COROLLARY. $\exists$ a bijection $f: A \rightarrow B \Leftrightarrow \exists$ a bijection $g: B \rightarrow A$.

As shown by the following theorem, we can obtain from every injection a bijection.

THEOREM 2.4. *If $f: A \rightarrow B$ is an injection, then $f: A \rightarrow f(A)$ is a bijection.*

Proof. By assumption $f \subset A \times B$. But as $\langle x, y \rangle \in f$ implies $y = f(x) \in f(A)$, we have $f \subset A \times f(A)$. Since for every x in A there exists exactly one y in $f(A)$ such that $\langle x, y \rangle \in f$, we can legitimately write $f: A \rightarrow f(A)$. It is now easy to see that the latter mapping is surjective. By assumption, its graph f is injective.

6. Composite mapping

Let $f: A \to B$ and $g: B \to C$. Then every x in A has an image $f(x)$ in B and $f(x)$ has the image $g(f(x))$ in C. Therefore we can assign to every x in A the element $g(f(x))$ in C. In this way we can define a mapping from A to C.

DEFINITION 2.6. If $f: A \to B$ and $g: B \to C$ are any mappings then we denote by

$$A \longrightarrow B \xrightarrow{g} C \quad \text{or} \quad gf: A \longrightarrow C$$

the mapping such that the image of every x in A is $g(f(x))$. This mapping is called the *composite* of the two mappings $f: A \to B$ and $g: B \to C$.

THEOREM 2.5. *If $f: A \to B$ is a bijection and $f^{-1}: B \to A$ is the inverse bijection then*

(i) $$A \xrightarrow{f} B \xrightarrow{f^{-1}} A = A \xrightarrow{\iota_A} A,$$

(ii) $$B \xrightarrow{f^{-1}} A \xrightarrow{f} B = B \xrightarrow{\iota_B} B.$$

Proof. We use the equivalence of Theorem 2.3

$$\forall x \in B \text{ and } y \in A: f^{-1}(x) = y \Leftrightarrow f(y) = x.$$

We take an arbitrary $y \in A$ and denote $f(y)$ by x. Then, by the above equivalence, we have $f^{-1}(x) = y$ and if we substitute here $f(y)$ for x we obtain

$$f^{-1}(f(y)) = y = \iota_A(y).$$

This proves (i). Taking an arbitrary x in B and denoting $f^{-1}(x)$ by y, we have by the above equivalence that $f(y) = x$. Substituting here $f^{-1}(x)$ for y we obtain

$$f(f^{-1}(x)) = x = \iota_B(x)$$

which proves (ii).

EXERCISES

1. Let $f: A \to B$, $g: B \to C$, $h: C \to D$ be any mappings. Show that $h(gf): A \to D$ and $(hg)f: A \to D$ are the same mapping. (Thus we can use the notation $hgf: A \to D$ in place of either.)

2. Let $A = \{1, 2, 3\}$ and let $f: A \to A$, $g: A \to A$ be the mappings whose graphs are

$$f = \{\langle 1, 2\rangle, \langle 2, 3\rangle, \langle 3, 1\rangle\} \qquad g = \{\langle 1, 2\rangle, \langle 2, 1\rangle, \langle 3, 3\rangle\}.$$

Find fg, gf, ff, fgf, and f^{-1}. Verify that $fgf = g$ and $ff = f^{-1}$.

3. (i) Let $f: A \to B$, $g: B \to C$ both be surjective. Prove that $gf: A \to C$ is surjective.

(ii) Let $f: A \to B$, $g: B \to C$ both be injective. Prove that $gf: A \to C$ is injective.

4. Let $f: N \to N$ be the function defined by $f(x) = x^2$. Prove that for any two distinct mappings $g: N \to N$, $h: N \to N$ we have $fg \neq fh$. Give an example where $g \neq h$ and $gf = hf$.

5. Let $f: N \to N$ be defined by

$$f(x) = \begin{cases} \frac{1}{2}x, & \text{if } x \text{ is even} \\ x, & \text{if } x \text{ is odd.} \end{cases}$$

Prove that for any two distinct mappings $g : N \to N$, $h : N \to N$ we have $gf \neq hf$. Give an example where $g \neq h$ and $fg = fh$.

6. Let $f: A \to B$, $g: B \to C$ be bijections. Prove that $gf: A \to C$ is a bijection and show that $(gf)^{-1} = f^{-1}g^{-1}$.

7. Give an example of a set A and mappings $f: A \to A$ and $g: A \to A$ such that

(i) $f: A \to A$ is an injection but not a surjection,
(ii) $g: A \to A$ is a surjection but not an injection.

8. Let k be an arbitrary fixed natural number and let N_k denote the set $\{n \in N \mid n \geqslant k\}$. Find a bijection $f: N \to N_k$.

9. Let E be the set of all even natural numbers. Find a bijection $f: N \to E$. Find also a bijection $g: N \to (N - E)$.

10. If $f: A \to B$ and $C \subset A$, then we define $f(C)$ by

$$f(C) = \{f(x) \in B | x \in C\},$$

that is $y \in f(C) \Leftrightarrow y = f(x)$ for some x in C. We call two subsets D_1, D_2 of a set D, complementary in D if $D_1 \cap D_2 = \varnothing$ and $D_1 \cup D_2 = D$.

Prove that if $f: A \to B$ is injective and $A_1, A_2 \subset A$ are complementary in A, then $f(A_1), f(A_2)$ are complementary subsets of $f(A)$.

11. Let $g: (-1, 1) \to R$ be defined by $g(x) = x/(1-|x|)$. Prove that this mapping is a bijection.

CHAPTER THREE

Cardinal Numbers

1. The postulate of cardinal numbers

We have denoted the number of elements of a finite set A by $|A|$. But $|A|$ has not been defined precisely—we have understood $|A|$ in an intuitive sense. What is $|A|$? It is certainly some object associated with the set A. Before discussing the nature of $|A|$ let us look at the meaning of an equality $|A| = |B|$. What does $|A| = |B|$ mean? Suppose we have two finite sets A, B and wish to prove that they have the same number of elements each. Then we may count the number of elements in each, compare the results and see whether they are equal. But counting is not necessary. It is easy to see that the set A of fingers on the left hand has exactly as many elements as the set B of fingers on the right hand. All one has to do is to let every finger of the left hand touch the opposite finger of the right hand, thumb to thumb and so on. And then we can 'see' that the 'number' of fingers on both hands is the same, or alternatively what we can see is the mapping $f: A \to B$ which associates with every finger in A the finger in B which it meets. This mapping is obviously both injective and surjective, hence bijective.

Let us take another example: we have two boxes of matches, red matches in one, white matches in the other. We then can take out two matches at a time, one from each box, and if both boxes become empty simultaneously then the number of red matches is equal to that of the white ones. Again we have established here a bijection $f: A \to B$; the image of every red match x, from box A, is the white match y, from B, which has been taken out simultaneously with x.

We are now quite certain that if A and B are finite sets then $|A| = |B|$ is equivalent to saying that there exists a bijection $f: A \to B$ (or there exists a bijection $g: B \to A$, which means the same). Following Cantor we will carry over this idea to infinite sets. This will give us a method of deciding when two infinite sets have the same number of elements. The number of elements of an arbitrary (finite or infinite) set A will be called *the cardinal number* (or the *cardinality*) of A, and denoted by $|A|$. But what exactly is the cardinal number of A? It is an object associated with the set A and two sets A, B have the same cardinal number if and only if there is a bijection $f: A \to B$.

From our assumptions hitherto it does not follow that cardinal numbers exist; we have to assume this.

POSTULATE OF CARDINAL NUMBERS: For every set A there exists an object $|A|$, called the cardinal number of A, such that for any two sets A, B we have

$$|A| = |B| \Leftrightarrow \exists \text{ a bijection } f: A \to B.$$

Some cardinal numbers have names of their own, for example it is customary to use (after Cantor) the Hebrew letter $\aleph$ (aleph) with the suffix 0—written $\aleph_0$ (read aleph: zero) to denote $|N|$. $|R|$ is usually denoted by c and is called *the continuum*. We also denote $|\varnothing|$ by 0 ($\varnothing$ = the empty set) and we identify the cardinal number of the set $\{0, 1, 2, \ldots, n-1\}$ with n, so that

$$|\varnothing| = 0, \quad |\{0\}| = 1, \quad |\{0, 1\}| = 2, \quad |\{0, 1, 2\}| = 3,$$
$$|\{0, 1, 2, 3\}| = 4, \ldots$$

We may point out here that the above equalities could be used to define natural numbers 0, 1, 2, 3, 4,... in terms of the notions of 'cardinal number' and 'the empty set'.

2. The cardinal numbers $\aleph_0$ and c

The two cardinal numbers $\aleph_0$ and c are not equal to each other. To prove this fundamental fact, discovered by Cantor, we make the following digression on decimal expansions of real numbers.

Let a_0 be an integer and let $a_1, a_2, a_3, \ldots$ be an infinite sequence of digits, that is

$$a_n \in \{0, 1, 2, 3, 4, 5, 6, 7, 8, 9\} \quad \text{for } n \in \{1, 2, 3, \ldots\}.$$

Then the series $\sum_{n=0}^{\infty} \frac{a_n}{10^n}$ converges $\left(\text{it is majorized by } a_0 + \sum_{n=1}^{\infty} \frac{9}{10^n}\right)$. If x is the sum of this series, we shall denote x alternatively by $a_0 \cdot a_1 a_2 a_3 \ldots$. Hence in this new notation we have the equivalence

$$x = \sum_{n=0}^{\infty} \frac{a_n}{10^n} \Leftrightarrow x = a_0 . a_1 a_2 a_3 \ldots$$

where the a_n are integers such that $0 \leqslant a_n \leqslant 9$ when $n \geqslant 1$. We shall call the equality $x = a_0 \cdot a_1 a_2 a_3 \ldots$ a *decimal expansion* of x. For example,

$$\frac{1}{3} = 0 \cdot 3333\ldots, \qquad \frac{1}{5} = 0 \cdot 2000\ldots, \qquad \frac{1}{5} = 0 \cdot 1999\ldots$$

are decimal expansions of $x = 1/3$ and $1/5$. The expansion $x = a_0 \cdot a_1 a_2 a_3 \ldots$ will be called *proper* if $a_n \neq 9$ holds for infinitely many n. Thus the last expansion above is not proper, the first two are proper. We shall assume now without proof the following.

LEMMA. *Every real number x has a unique proper decimal expansion*

$$x = a_0 \cdot a_1 a_2 a_3 \ldots$$

THEOREM 3.1. $\aleph_0 \neq c$.

Proof. We have to show that $|N| = |R|$ is false. From the postulate of cardinal numbers we have

$$|N| \neq |R| \Leftrightarrow \text{there does not exist a bijection } f: N \to R.$$

Let us assume that there exists a bijection $f: N \to R$; we shall deduce a contradiction. By assumption $f: N \to R$ is surjective, thus

$$f(N) = \{f(0), f(1), f(2), \ldots\} = R.$$

Let us write the proper decimal expansions of the numbers $f(n)$

$$f(0) = a_0^{(0)} \cdot a_1^{(0)} a_2^{(0)} a_3^{(0)} \ldots a_n^{(0)} \ldots$$
$$f(1) = a_0^{(1)} \cdot a_1^{(1)} a_2^{(1)} a_3^{(1)} \ldots a_n^{(1)} \ldots$$
$$f(2) = a_0^{(2)} \cdot a_1^{(2)} a_2^{(2)} a_3^{(2)} \ldots a_n^{(2)} \ldots$$
$$\cdots\cdots\cdots\cdots\cdots\cdots$$
$$f(n) = a_0^{(n)} \cdot a_1^{(n)} a_2^{(n)} a_3^{(n)} \ldots a_n^{(n)} \ldots$$
$$\cdots\cdots\cdots\cdots\cdots\cdots$$

We take now an arbitrary integer $b_0 \neq a_0^{(0)}$ and any numbers $b_k \in N$ such that

$$a_k^{(k)} \neq b_k < 9 \quad \text{for} \quad k = 1, 2, 3, \ldots$$

that is $a_1^{(1)} \neq b_1 < 9$, $a_2^{(2)} \neq b_2 < 9, \ldots$ and so on. (We have at least 8 possible choices for each b_k.) Let $x = \sum_{n=0}^{\infty} \frac{b_n}{10^n}$. Then

$$x = b_0 \cdot b_1 b_2 b_3 \ldots$$

is a decimal expansion of x, and this expansion is proper by our choice of the b_k. We have $x \in R$ and thus there must be some $n \in N$ such that $x = f(n)$, whence

$$x = b_0 \cdot b_1 b_2 b_3 \ldots b_n \ldots = a_0^{(n)} \cdot a_1^{(n)} a_2^{(n)} a_3^{(n)} \ldots a_n^{(n)} \ldots = f(n).$$

Since both decimal expansions appearing above are proper expansions of the same number x $(= f(n))$, the lemma stated just before the theorem implies that these expansions are identical, that is

$$b_0 = a_0^{(n)}, \qquad b_1 = a_1^{(n)}, \qquad b_2 = a_2^{(n)}, \qquad \ldots, \qquad b_n = a_n^{(n)},$$
$$b_{n+1} = a_{n+1}^{(n)}, \ldots$$

Among these equalities we have obtained $b_n = a_n^{(n)}$, and this contradicts our general rule $a_k^{(k)} \neq b_k < 9$ used in choosing b_n. Thus $\aleph_0 \neq c$.

Let us consider the set $N \times N$. We have $N \times N \subset R \times R$ since $N \subset R$ and as we called $R \times R$ the plane, $N \times N$ is the set of those points $\langle x, y \rangle$ in the plane whose coordinates x and y are natural numbers (Figure 7). For any fixed k in N let A_k be the subset of $N \times N$ defined by

$$A_k = \{\langle 0, k \rangle, \langle 1, k \rangle, \langle 2, k \rangle, \ldots, \langle n, k \rangle, \langle n+1, k \rangle, \ldots\}$$

that is, A_k consists of all those points of $N \times N$ whose y-coordinates are equal to k. Then $|A_k| = \aleph_0$ because there exists a bijection $f: A_k \to N$, for example we can define $f(\langle x, k \rangle) = x$, for every x. As no two of the sets $A_0, A_1, A_2, \ldots$ have any elements in

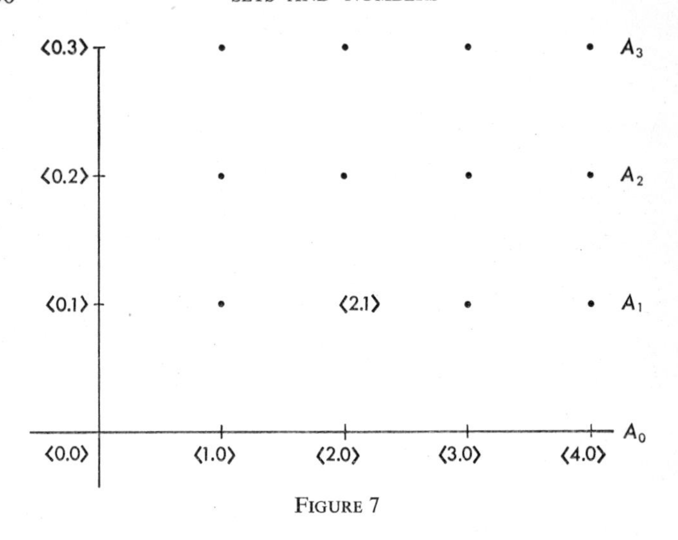

FIGURE 7

common one might expect that $\bigcup_{k=0}^{\infty} A$ would have a different cardinal number (in some sense 'greater') than each A_k—different from $\aleph_0$. But it is evident that $\bigcup_{k=0}^{\infty} A_k = N \times N$, thus from the theorem below it will follow that $|\bigcup_{k=0}^{\infty} A_k| = \aleph_0$.

THEOREM 3.2. $|N \times N| = \aleph_0$.

Proof. It is enough to find a bijection $g: N \to N \times N$. Such a bijection can be defined as follows. Let n be any natural number. Let k denote the greatest natural number such that $n+1$ is divisible by 2^k (if $n+1$ is odd then $k = 0$). Thus $n+1 =$

$2^k m$, where m is odd. The number m can be written as an even number plus one, and every even number is of the form $2l$, where $l \in N$. So we have finally

$$n+1 = 2^k(2l+1).$$

In this way we have associated with n an element $\langle k, l\rangle$ in $N\times N$. We now define a mapping $g: N \to N\times N$ by putting $g(n) = \langle k, l\rangle$. Then g is a surjection because every $\langle k, l\rangle$ in $N\times N$ is an image: $\langle k, l\rangle = g(2^k(2l+1)-1)$. But g is also an injection, for if we have distinct numbers n_1, n_2 in N and $n_1+1 = 2^{k_1}(2l_1+1)$, $n_2+1 = 2^{k_2}(2l_2+1)$, then we cannot have $k_1 = k_2$ and $l_1 = l_2$, in other words, we cannot have $\langle k_1, l_1\rangle = \langle k_2, l_2\rangle$, for then $n_1 \neq n_2$. This proves that $g(n_1) \neq g(n_2)$. The proof is complete.

The cardinal number of a set tells us 'how large' the set is but if we measure the 'size' of a set in some other way, different information may result—for example, there are large and small intervals in R if we compare them by length, but we shall prove that all the intervals have the same cardinal number. We prove first the theorem

THEOREM 3.3. *Any two open intervals have the same cardinal number.*

Proof. Let (a, b) and (d, e) be open intervals where $a < b$ and $d < e$. Then a bijection $f: (a, b) \to (d, e)$ is given by

$$f(x) = \frac{x-a}{b-a}(e-d)+d.$$

Indeed, it is easy to see that $a < x < b \Rightarrow d < f(x) < e$, so that the above formula really defines a mapping $f: (a, b) \to (d, e)$.

We have $x_1 < x_2 \Rightarrow f(x_1) < f(x_2)$ for any x_1, x_2 and thus the mapping is injective. It is also surjective, for if $y \in (d, e)$, then $y = f(x)$ for $x = \dfrac{y-d}{e-d}(b-a)+a$.

THEOREM 3.4. *Every open interval has cardinality c.*

Proof. We have to show that $|(a, b)| = |R|$, that is, we should find a bijective mapping $f: (a, b) \to R$. According to the previous theorem it does not matter here which open interval we consider; let us consider the interval $(-1, 1)$. Then the function

$$g: (-1, 1) \to R, \quad \text{where} \quad g(x) = \frac{x}{1-|x|}$$

is a bijection. (See Exercise 11, p. 24).

Another example of a bijection from an open interval to R is the mapping

$$f: \left(-\frac{1}{2}\pi, \frac{1}{2}\pi\right) \to R, \quad \text{where} \quad f(x) = \tan x.$$

Notation. We shall denote arbitrary cardinal numbers by letters of the Greek alphabet: $\alpha, \beta, \gamma, \ldots$.

3. Inequality

We wish to define the meaning of $\alpha < \beta$; it is simpler to give first the definition of $\alpha \leqslant \beta$. So if A, B are sets satisfying $|A| = \alpha$, $|B| = \beta$, we should like to define when B has 'at least as many' elements as A. Suppose that there exists an injection $f: A \to B$.

Then intuitively B has 'at least as many' elements as A, since B is 'spacious' enough to accommodate all the distinct images of the distinct elements of A. We shall make this a basis of our definition.

DEFINITION 3.1. Let α, β be cardinal numbers and let A, B be sets such that $|A| = \alpha$, $|B| = \beta$. We define $\alpha \leqslant \beta$ by

$$\alpha \leqslant \beta \Leftrightarrow \exists \text{ injection } f: A \to B.$$

Now this definition might have no meaning; as it stands it might be nonsense. For suppose that we find some other two sets A_1, B_1 such that $|A_1| = \alpha$, $|B_1| = \beta$ but there does not exist an injection $f_1 : A_1 \to B_1$. Then we should not write $\alpha \leqslant \beta$; on the other hand the above sets A, B indicate that we may write $\alpha \leqslant \beta$. We can save the definition only if we prove that such 'bad' sets A_1, B_1 cannot exist—so we must prove:

THEOREM 3.5. *If* $|A_1| = |A|$, $|B_1| = |B|$ *then*

$$\exists \textit{ an injection } f: A \to B \Leftrightarrow \exists \textit{ an injection } f_1 : A_1 \to B_1.$$

Proof. From $|A_1| = |A|$ and $|B_1| = |B|$ there follows the existence of two bijections $h : A \to A_1$, $k : B \to B_1$ as shown in Figure 8.
Now suppose that there exists an injection $f: A \to B$; then we denote by f_1 the mapping $kfh^{-1} : A_1 \to B_1$. This is injective because it is a composite of injections (see Exercise 3(ii), p. 23), and so we have shown that f_1 exists. Conversely, if f_1 exists we denote by f the mapping $k^{-1}f_1h : A \to B$ which is again injective. This proves the theorem.

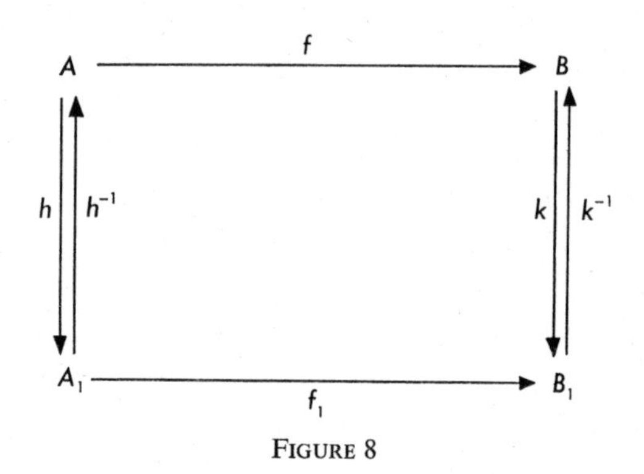

FIGURE 8

THEOREM 3.6. *If A, B are any two sets, then we have*

$$|A| \leqslant |B| \Leftrightarrow \exists \quad \text{an injection} \quad f: A \to B.$$

Proof. Denote $|A|$ and $|B|$ by α and β, then use Definition 3.1.

A consequence of Theorem 3.6 is:

THEOREM 3.7. $A \subset B \Rightarrow |A| \leqslant |B|$.

Proof. In this case an injection $f: A \to B$ is given by $f(x) = x$ for all x in A.

A fundamental property of inequality is given in the following:

THEOREM 3.8. (Transitivity of inequality). $\alpha \leqslant \beta$ *and* $\beta \leqslant \gamma \Rightarrow \alpha \leqslant \gamma$.

Proof. Let A, B, C be sets such that $|A| = \alpha$, $|B| = \beta$, and $|C| = \gamma$. By assumption there exist injections $f: A \to B$ and

$g : B \to C$. Their composite

$$gf : A \to C$$

is an injection by Exercise 3(ii), p. 23. Hence $|A| \leqslant |C|$, that is, $\alpha \leqslant \gamma$.

DEFINITION 3.2. $\alpha < \beta \Leftrightarrow \alpha \leqslant \beta$ and $\alpha \neq \beta$.

THEOREM 3.9. $\aleph_0 < c$.

Proof. We have $N \subset R$, hence $|N| \leqslant |R|$, that is, $\aleph_0 \leqslant c$. We have shown previously that $\aleph_0 \neq c$, hence $\aleph_0 < c$.

4. Addition

If we have two natural numbers m, n, then their sum $m+n$ can be obtained as follows. We take two disjoint sets A, B such that $|A| = m$, $|B| = n$ and then $m+n = |A \cup B|$. We generalize this to arbitrary cardinal numbers as follows:

DEFINITION 3.3. We shall denote by $\alpha+\beta$ the cardinal number of $|A \cup B|$ where A, B are any sets such that $A \cap B = \varnothing$ and $|A| = \alpha$, $|B| = \beta$.

This definition is incomplete: we still have to verify that (i) for any α, β there exist disjoint sets A, B such that $|A| = \alpha$, $|B| = \beta$, (ii) $|A \cup B|$ does not depend on the choice of the sets A, B satisfying condition (i).

Concerning (i), it is clear that there are sets A_1, B_1 such that $|A_1| = \alpha$, $|B_1| = \beta$, since a cardinal number exists only in association with a set. Now take two distinct objects—for example, the letters a, b and define

$$A = \{\langle x, a\rangle \mid x \in A_1\} \quad \text{and} \quad B = \{\langle x, b\rangle \mid x \in B_1\}.$$

The sets A, B are disjoint ($\langle x, a\rangle$ cannot belong to B because $a \neq b$) and they have the required cardinal numbers as well ($\exists$ bijections $f: A \to A_1$, $g: B \to B_1$).

Condition (ii) follows from

THEOREM 3.10. *If A, B, A', B' are sets such that $|A| = |A'|$, $|B| = |B'|$, and $A \cap B = \varnothing$, $A' \cap B' = \varnothing$, then $|A \cup B| = |A' \cup B'|$.*

Proof. By assumption we have bijections $f: A \to A'$, $g: B \to B'$. Let us denote by $h: (A \cup B) \to (A' \cup B')$ the mapping which is equal to f on A and to g on B (if $x \in A$ then $h(x) = f(x)$, and if $x \in B$ then $h(x) = g(x)$). Then it is clear that $h: (A \cup B) \to (A' \cup B')$ is a bijection, whence $|A \cup B| = |A' \cup B'|$.

THEOREM 3.11. *If A and B are any sets such that $A \cap B = \varnothing$, then*

$$|A| + |B| = |A \cup B|.$$

Proof. Denote $|A|$ and $|B|$ by α and β, then use Definition 3.3.

If A, B are disjoint sets, both of cardinality $\aleph_0$, then their union is of cardinality $\aleph_0$. In other words, we have

THEOREM 3.12. $\aleph_0 + \aleph_0 = \aleph_0$.

Proof. Take $A = \{1, 3, 5, \ldots\}$, $B = \{0, 2, 4, 6, \ldots\}$. Then $|A| = \aleph_0$ and $|B| = \aleph_0$ (see Exercise 9 on p. 24); hence

$$\aleph_0 + \aleph_0 = |A| + |B| = |A \cup B| = |N| = \aleph_0.$$

The integers are the numbers $0, \pm 1, \pm 2, \ldots$. Let us denote by Z the set of all integers. We shall deduce from the previous theorem that

THEOREM 3.13. $|Z| = \aleph_0$.

Proof. Let N^- denote the set $\{0, -1, -2, -3, \ldots\}$. Then $|N^-| = \aleph_0$ because we have the bijection $f: N \to N^-$ defined by $f(x) = -x$ for all x in N. Let $N_1 = \{1, 2, 3, \ldots\}$, then $|N_1| = \aleph_0$ (see Exercise 8 p. 24). Finally

$$|Z| = |N_1 \cup N^-| = |N_1| + |N^-| = \aleph_0 + \aleph_0 = \aleph_0.$$

THEOREM 3.14. *If* $n \in N$, *then* $n + \aleph_0 = \aleph_0$.

Proof. We have $|\{n, n+1, n+2, \ldots\}| = \aleph_0$ (see Exercise 8, p. 24) and $|\{0, 1, \ldots, n-1\}| = n$.
Hence,

$$n + \aleph_0 = |\{0, 1, \ldots, n-1\} \cup \{n, n+1, n+2, \ldots\}| = |N| = \aleph_0.$$

THEOREM 3.15 (Commutativity of addition). $\alpha + \beta = \beta + \alpha$.

Proof. Take A, B such that $A \cap B = \varnothing$ and $|A| = \alpha$, $|B| = \beta$. Then $\alpha + \beta = |A \cup B| = |B \cup A| = \beta + \alpha$.

Inequality can be defined entirely in terms of addition, for we have

THEOREM 3.16. $\alpha \leqslant \beta \Leftrightarrow \exists \gamma$ s.t. $\alpha + \gamma = \beta$.

Proof. To prove the implication $\Leftarrow$, assume that $\alpha + \gamma = \beta$ and take sets A, C, s.t. $|A| = \alpha$, $|C| = \gamma$ and $A \cap C = \varnothing$. Then

$|A \cup C| = \beta$. We have $A \subset A \cup C$, whence $|A| \leqslant |A \cup C|$, that is, $\alpha \leqslant \beta$. To prove the implication $\Rightarrow$, we take two sets A, B such that $|A| = \alpha$, $|B| = \beta$ and we assume that $\alpha \leqslant \beta$. Then there exists an injection $f: A \to B$ which, by Theorem 2.4 on p. 21, implies the existence of the bijection $f: A \to f(A)$. Thus $|f(A)| = |A| = \alpha$. Now write

$$B = f(A) \cup (B - f(A));$$

B is thus the union of two disjoint sets and hence

$$\beta = |B| = |f(A)| + |B - f(A)| = \alpha + \gamma$$

where by γ we have denoted $|B - f(A)|$.

In calculating cardinal numbers, the following property of addition is very useful

THEOREM 3.17. (Associativity of addition).

$$(\alpha + \beta) + \gamma = \alpha + (\beta + \gamma).$$

Proof. Let us note first that there exist three sets A, B, C such that $A \cap B = \varnothing$, $B \cap C = \varnothing$, $C \cap A = \varnothing$ (such sets are called *pairwise disjoint*) and moreover $|A| = \alpha$, $|B| = \beta$ and $|C| = \gamma$. There certainly exist sets A_1, B_1, C_1 such that $|A_1| = \alpha$, $|B_1| = \beta$ and $|C_1| = \gamma$. It is enough now to take three distinct objects— for example the letters a, b, c— and define

$$A = \{\langle x, a\rangle \mid x \in A_1\}, \qquad B = \{\langle x, b\rangle \mid x \in B_1\},$$
$$C = \{\langle x, c\rangle \mid x \in C_1\}.$$

We have

$$(\alpha+\beta)+\gamma = |A \cup B|+|C| = |(A \cup B) \cup C| = |A \cup (B \cup C)|$$
$$= |A|+|B \cup C| = \alpha+(\beta+\gamma).$$

The preceding three theorems imply

THEOREM 3.18. (Invariance of inequality with respect to addition).

$$\alpha \leqslant \beta \Rightarrow \alpha+\delta \leqslant \beta+\delta \quad \textit{for every } \delta.$$

Proof. By Theorems 3.15 and 3.16, $\alpha \leqslant \beta \Rightarrow \exists \gamma$ s.t. $\gamma+\alpha = \beta$. Hence, by Theorems 3.15 and 3.17

$$(\alpha+\delta)+\gamma = \gamma+(\alpha+\delta) = (\gamma+\alpha)+\delta = \beta+\delta.$$

Using again Theorem 3.16 we get $\alpha+\delta \leqslant \beta+\delta$.

5. Infinite sets

We should now like to ask questions such as: If A is an infinite set and B is finite, is $|A \cup B| = |A|$? And if we have equality here, will it still be true after we take B to be of cardinality $\aleph_0$? To answer such questions we need to have a precise definition of an infinite cardinal number. We shall accept the following

DEFINITION 3.4. The cardinal number α is *infinite* $\Leftrightarrow$ $\aleph_0 \leqslant \alpha$.

Accordingly, we shall call a set A infinite if $\aleph_0 \leqslant |A|$. The answers to the questions raised above are given by the following two theorems.

THEOREM 3.19. *α infinite* $\Rightarrow \alpha + \aleph_0 = \alpha$.

Proof. α infinite $\Rightarrow \aleph_0 \leqslant \alpha \Rightarrow \exists \gamma$ s.t. $\gamma + \aleph_0 = \alpha$, by Theorems 3.16 and 3.15. Thus $\alpha + \aleph_0 = (\gamma + \aleph_0) + \aleph_0 = \gamma + (\aleph_0 + \aleph_0) = \gamma + \aleph_0 = \alpha$.

THEOREM 3.20. *α infinite* $\Rightarrow \alpha + n = \alpha$ *for every* $n \in N$.

Proof. α infinite $\Rightarrow \alpha + \aleph_0 = \alpha$. Hence $\alpha + n = (\alpha + \aleph_0) + n = \alpha + (\aleph_0 + n) = \alpha + \aleph_0 = \alpha$, since we have proved that $\aleph_0 + n = \aleph_0$ (Theorems 3.14 and 3.15).

THEOREM 3.21. *If $a < b$, then the closed interval $[a, b]$ has cardinal number c.*

Proof. Using Theorem 3.20 we get $|[a, b]| = |(a, b) \cup \{a, b\}| = |(a, b)| + |\{a, b\}| = c + 2 = c$.

According to our definition, a set A is infinite if $\aleph_0 \leqslant |A|$, that is, if there exists an injection $f : N \rightarrow A$. It is natural to ask whether infinite sets can be characterized by a simple intrinsic property which does not involve the very special set N. To state the property in question, let us call a subset $B \subset A$ *proper* if $B \neq A$. Then we have

THEOREM 3.22. *A is infinite* $\Leftrightarrow \exists$ *a proper subset* $B \subset A$ s.t. $|A| = |B|$.

Proof. To prove the implication $\Rightarrow$ assume that A is infinite. Then by Definition 3.4 there exists an injection $f : N \rightarrow A$ and

hence $A = f(N) \cap (A - f(N))$. We have $f(N) = \{f(0), f(1), f(2), \ldots\}$ and since $f: N \to f(N)$ is a bijection (see Theorem 2.4 on p. 21), $|f(N)| = \aleph_0$. Therefore

$$|A| = \aleph_0 + |A - f(N)|.$$

Now let B_1 be the set $\{f(1), f(2), f(3), \ldots\}$. Then $|B_1| = \aleph_0$ since there is a bijection $g: N \to B_1$ defined by $g(n) = f(n+1)$ for all n. Let $B = B_1 \cup (A - f(N))$. Then B is a proper subset of A since the element $f(0)$ of A belongs neither to B_1 nor to $A - f(N)$. And we have

$$|B| = |B_1| + |A - f(N)| = \aleph_0 + |A - f(N)| = |A|.$$

To prove the implication $\Leftarrow$ assume that B is a proper subset of A such that $|A| = |B|$. Then there exists a bijection $f: A \to B$, and since $B \subset A$, the mapping $f: A \to B$ assigns to every element x in A some element $f(x)$ in A. Since this mapping is a surjection, we have $f(A) = B$, and since $B \subset A$ is proper, we have $A - f(A) \neq \varnothing$. Hence there exists some $x_0 \in A - f(A)$. Consider now the sequence $x_0, x_1, x_2, x_3, \ldots$ of elements of A defined by

$$x_1 = f(x_0), \qquad x_2 = f(x_1), \qquad x_3 = f(x_2), \qquad \ldots,$$
$$x_{n+1} = f(x_n), \ldots$$

Let us prove that

$$k \neq l \Rightarrow x_k \neq x_l \quad \text{for every} \quad k, l \text{ in } N.$$

For, assume to the contrary that $x_k = x_l$ and $k \neq l$, say $k < l$. If $k > 0$, then we have from the definition of x_k that $x_k = f(x_{k-1})$ and also $x_l = f(x_{l-1})$. Thus $f(x_{k-1}) = f(x_{l-1})$ and since f is injective, it follows that $x_{k-1} = x_{l-1}$. If still $k-1 > 0$, we

write $x_{k-1} = f(x_{k-2})$, $x_{l-1} = f(x_{l-2})$ and we deduce that $x_{k-2} = x_{l-2}$. If still $k-2 > 0$, then we apply again the above argument, and it is clear that after k steps altogether we arrive at the equality $x_0 = x_{k-k} = x_{l-k}$. But $l-k > 0$, whence $x_{l-k} = f(x_{l-k-1})$. If follows now that $x_0 = f(x_{l-k-1}) \in f(A)$ contrary to $x_0 \in A - f(A)$. This shows that we cannot have $x_k = x_l$ and $k \neq l$.

Now define a mapping $g : N \to A$ by putting $g(n) = x_n$ for every n in N. Then $g : N \to A$ is an injection, for if $k \neq l$, then, as we have just proved, $g(k) \neq g(l)$. Thus $\aleph_0 \leqslant |A|$, and therefore A is infinite. This completes our proof.

If we have two equivalent propositions, then their negations are also equivalent to each other. Applying this law to the equivalence in the preceding theorem, we obtain:

THEOREM 3.23. *A is finite* $\Leftrightarrow$ *for every proper subset B of A we have* $|B| \neq |A|$.

Here we can replace $|B| \neq |A|$ by $|B| < |A|$, since $B \subset A$ implies $|B| \leqslant |A|$.

6. The Cantor–Bernstein–Schröder theorem

Let R^+ denote the set of all positive real numbers. Then

$$(0, 1) \subset R^+ \subset R.$$

If we denote $|R^+|$ by α, then the above inclusions prove that $c \leqslant \alpha$ and $\alpha \leqslant c$. Can we deduce that $\alpha = c$? We can, as a consequence of the general theorem conjectured by Cantor and then proved by F. Bernstein and E. Schröder:

THEOREM 3.24. *For any cardinal numbers* α, β

$$\alpha \leqslant \beta \quad \textit{and} \quad \beta \leqslant \alpha \Rightarrow \alpha = \beta.$$

The proof will be preceded by two lemmas. We recall that if $f: A \to B$ and $C \subset A$ then $f(C)$ denotes the set of all images $f(x)$ of elements $x \in C$ (see Exercise 10, p. 24). We shall call $f(C)$ the image of C under $f: A \to B$.

LEMMA 1. *If* $f: A \to B$ *is injective and* $C \subset A$, *then* $|f(C)| = |C|$.

Proof. Let us first consider the mapping $g: C \to A$ where $g(x) = x$ for every x in C. Then $g: C \to A$ and $f: A \to B$ are injections whence $fg: C \to B$ is an injection (Exercise 3(ii), p. 23). Let us denote fg by h. Then the existence of an injection $h: C \to B$ implies the existence of the bijection $h: C \to h(C)$, by Theorem 2.4, p. 21. But it is clear that an element y of B is of the form $h(x)$, where $x \in C$, if and only if

$$y = f(g(x)) = f(x), \quad \text{where } x \in C,$$

and thus $h(C) = f(C)$. We have now a bijection $h: C \to f(C)$, and this proves that $|f(C)| = |C|$.

LEMMA 2. *If* $f: A \to B$ *is injective and* $C \subset A$, *then*

$$f(A) - f(C) = f(A - C).$$

Proof. We shall use the result of Exercise 10, p. 24. The sets C, $A - C$ are obviously complementary subsets of A. Hence $f(C)$ and $f(A - C)$ must be complementary subsets of $f(A)$, that is,

$f(C) \cap f(A-C) = \varnothing$ and $f(C) \cup f(A-C) = f(A)$. Hence $f(A) - f(C) = (f(C) \cup f(A-C)) - f(C) = f(A-C)$.

Proof of the Cantor–Bernstein–Schröder theorem. Let A, B be two sets such that $|A| = \alpha$ and $|B| = \beta$. By assumption there are injections $f: A \to B$ and $g: B \to A$. Consider the image $g(B)$ of B in A. If $g(B) = A$ then $g: B \to A$ is a bijection and hence $\alpha = \beta$. Assume now $g(B) \neq A$ and denote the set

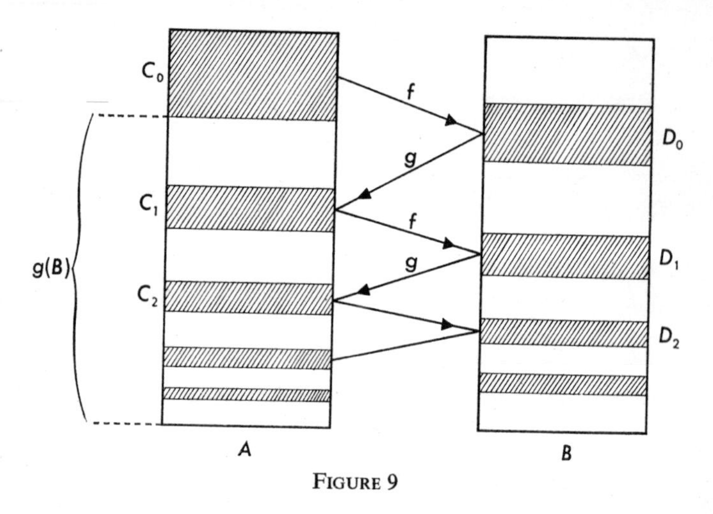

FIGURE 9

$A - g(B)$ by C_0 (see Figure 9). This set C_0 gives rise to two sequences of sets

$$C_0, C_1, C_2, \ldots, C_n, \ldots \subset A, \; D_0, D_1, D_2, \ldots, D_n, \ldots \subset B,$$

such that $D_0 = f(C_0)$, $C_1 = g(D_0)$, $D_1 = f(C_1)$, $C_2 = g(D_1)$, $\ldots$, $D_n = f(C_n)$, $C_{n+1} = g(D_n), \ldots$ (shaded sets in Figure 9).

Let $C = \bigcup_{k=0}^{\infty} C_k$ and $D = \bigcup_{k=0}^{\infty} D_k$.

Then C and D are the shaded parts of A and B respectively. It follows from our definition of $D_0, D_1, D_2, \ldots$ and $C_0, C_1, C_2, \ldots$ that $f(C) = D$. Therefore by Lemma 1 above, $|C| = |D|$. Now let us show that $|A-C| = |B-D|$. From the definition of $D_0, D_1, D_2, \ldots$ and $C_0, C_1, C_2, \ldots$ it follows that

$$g(D) = \bigcup_{k=1}^{\infty} C_k.$$

On the other hand from the definition of C_0, we have $g(B) = A - C_0$. Hence, using Lemma 2,

$$g(B-D) = g(B) - g(D) = (A - C_0) - \bigcup_{k=1}^{\infty} C_k = A - \bigcup_{k=0}^{\infty} C_k = A - C.$$

By Lemma 1, this proves that $|A-C| = |B-D|$. Knowing already that $|C| = |D|$, we now can write

$$\alpha = |A| = |(A-C) \cup C| = |A-C| + |C| = |B-D| + |D| =$$
$$= |(B-D) \cup D| = |B| = \beta$$

which we wished to show.

Let us denote the set of all rational numbers by Q. We shall use the above theorem to find the cardinal number of Q.

THEOREM 3.25. $|Q| = \aleph_0$.

Proof. Let Q^+ denote the set of all positive rational numbers. We assume every $x \in Q^+$ written in the form $x = k/n$, where $k, n \in N$ are both different from 0 and relatively prime (every positive rational number can be written in this form in precisely one way). We shall show first that $|Q^+| = \aleph_0$.

Let $f: N \to Q^+$ be given by $f(x) = x+1$ for every x in N. Clearly this mapping is injective, whence $\aleph_0 \leqslant |Q^+|$.

Now consider the mapping $g: Q^+ \to N \times N$ given by $g(k/n) = \langle k, n \rangle$. It is clear that

$$k_1/n_1 \neq k_2/n_2 \Rightarrow \langle k_1, n_1 \rangle \neq \langle k_2, n_2 \rangle,$$

that is, g is injective. This proves that $|Q^+| \leqslant |N \times N|$. But we have shown (Theorem 3.2, p. 30) that $|N \times N| = \aleph_0$. Therefore $|Q^+| \leqslant \aleph_0$ which together with $\aleph_0 \leqslant |Q^+|$ implies, by the CBS-theorem, that $|Q^+| = \aleph_0$.

Now we can find the cardinal number of Q. We denote the set of all negative rational numbers by Q^-. Then $|Q^-| = |Q^+|$ since we have a bijection $h: Q^- \to Q^+$ given by $h(x) = -x$. Obviously

$$Q = Q^+ \cup \{0\} \cup Q^-,$$

whence, using Theorems 3.12 and 3.14,

$$\begin{aligned} |Q| &= |Q^+| + |\{0\} \cup Q^-| = |Q^+| + (1 + |Q^-|) \\ &= \aleph_0 + (1 + \aleph_0) = \aleph_0 + \aleph_0 = \aleph_0. \end{aligned}$$

EXERCISES

1. Let A, B be any two sets, such that $|A| = |B| = \aleph_0$. Prove that $|A \cup B| = \aleph_0$.

2. Prove that every subset of R which contains a non-empty open interval has cardinality c. Deduce that $c + c = c$.

3. Prove that the set $R-Q$ (the set of irrational numbers) has cardinality c.

4. If A, B are any sets, prove that $|A|+|B| = |A\cup B|+|A\cap B|$.

5. Let A, B be such that $\aleph_0 \leqslant |A|$ and $|B| \leqslant \aleph_0$. Prove that $|A\cup B| = |A|$.

6. Prove that if $\gamma+\alpha = \alpha$ and $\alpha \leqslant \beta$, then $\gamma+\beta = \beta$.

CHAPTER FOUR

Cardinal Algebra

The operations of addition $\alpha+\beta$, multiplication $\alpha\beta$ and exponentiation α^β can be defined for arbitrary cardinal numbers α, β; moreover if α, β are finite (if $\alpha, \beta \in N$), then these operations have the usual meaning. We have already seen this for $\alpha+\beta$. In this chapter we shall discuss the operations $\alpha\beta$ and α^β. The latter is of particular interest for it leads to cardinal numbers greater than c. These operations are useful tools in finding cardinal numbers of various known sets.

1. Product

If m and n are natural numbers, then a way to obtain their product mn is the following: We take two sets A, B such that $|A| = m$ and $|B| = n$; then $mn = |A \times B|$ (see Theorem 2.1, p. 14). We accept now

DEFINITION 4.1. If α, β are any cardinal numbers, and A, B are sets such that $|A| = \alpha$, $|B| = \beta$, then we define $\alpha\beta$ by

$$\alpha\beta = |A \times B|.$$

We have to show that $\alpha\beta$ is well defined—that is, we have to prove that $|A \times B|$ depends only on $|A|$ and $|B|$.

THEOREM 4.1.

$$|A_1| = |A| \text{ and } |B_1| = |B| \Rightarrow |A_1 \times B_1| = |A \times B|.$$

Proof. We have bijections $f: A_1 \to A$, $g: B_1 \to B$. Consider the mapping $h:(A_1 \times B_1) \to (A \times B)$ defined by $h(\langle x, y \rangle) = \langle f(x), g(y) \rangle$ for every $\langle x, y \rangle$ in $A_1 \times B_1$. To prove that h is injective, assume that $\langle x', y' \rangle \neq \langle x, y \rangle$. Then $x' \neq x$ or $y' \neq y$, whence $f(x') \neq f(x)$ or $g(y') \neq g(y)$ since f and g are injective. Therefore $\langle f(x'), g(y') \rangle \neq \langle f(x), g(y) \rangle$. To prove that $h: A_1 \times B_1 \to A \times B$ is a surjection, take an arbitrary element $\langle a, b \rangle$ of $A \times B$. Then $a \in A$ and therefore $a = f(x)$ for some x in A_1, since $f: A_1 \to A$ is surjective. Similarly $b \in B$ is of the form $g(y)$ where $y \in B_1$. Hence

$$\langle a, b \rangle = \langle f(x), g(y) \rangle = h\langle x, y \rangle.$$

THEOREM 4.2. *If A, B are any sets, then* $|A|.|B| = |A \times B|$.

Proof. Denote $|A|$ and $|B|$ by α and β, then use Definition 4.1.

THEOREM 4.3. $\aleph_0 \aleph_0 = \aleph_0$.

Proof. $\aleph_0 \aleph_0 = |N|.|N| = |N \times N| = \aleph_0$, by Theorem 3.2, p. 30.

We have called $R \times R$ 'the plane'. Similarly, if I denotes the open interval (0, 1), we shall call $I \times I$ the open square (each point of $I \times I$ is a pair of real numbers, both between 0 and 1). Thus $|I \times I|$ is the number of points inside the square. We shall prove now that $|I \times I| = |I|$ which is a surprising fact: There are exactly as many points in a square as in an interval.

THEOREM 4.4. $|I \times I| = |I|$.

Proof. It is easy to see that $|I| \leqslant |I \times I|$, for we have an injection $f : I \to I \times I$ given by $f(x) = \langle x, \frac{1}{2} \rangle$. To prove that $|I \times I| \leqslant |I|$, we shall find an injection $g : I \times I \to I$. This mapping will assign to every $\langle x, y \rangle$ in $I \times I$ a number $z = g\langle x, y \rangle$ in I defined as follows. We write the proper decimal expansions of x and y

$$x = 0 \cdot a_1 a_2 a_3 \ldots a_n \ldots \qquad (0 \leqslant a_n \leqslant 9)$$
$$y = 0 \cdot b_1 b_2 b_3 \ldots b_n \ldots \qquad (0 \leqslant b_n \leqslant 9)$$

(see p. 27) and we define the number $z = g\langle x, y \rangle$ to be the number whose decimal expansion is

$$z = 0 \cdot a_1 b_1 a_2 b_2 a_3 b_3 \ldots a_n b_n \ldots .$$

This expansion is certainly proper. To show that the mapping $g : I \times I \to I$ is injective, consider $\langle x', y' \rangle \neq \langle x, y \rangle$. Then $x' \neq x$ or $y' \neq y$. If $x' \neq x$, then the proper decimal expansion of x' differs from that of x at some digit, and therefore the expansion of

$$z' = g(\langle x', y' \rangle) = 0.a_1' b_1' a_2' b_2' a_3' b_3' \ldots a_n' b_n' \ldots$$

will be different from the expansion of z. This implies $z' \neq z$ (since one number z cannot have two distinct proper expansions). So we have

$$x' \neq x \Rightarrow g(\langle x', y' \rangle) \neq g\langle x, y \rangle.$$

Similarly we show that $y' \neq y \Rightarrow g(\langle x', y' \rangle) \neq g(\langle x, y \rangle)$ so that finally we have

$$\langle x', y' \rangle \neq \langle x, y \rangle \Rightarrow g(\langle x', y' \rangle) \neq g(\langle x, y \rangle).$$

From $|I| \leqslant |I \times I|$ and $|I \times I| \leqslant |I|$ we deduce equality, by the CBS-theorem.

THEOREM 4.5. $cc = c$.

Proof. We have $cc = |I|.|I| = |I \times I| = |I| = c$, by Theorems 4.2 and 4.4.

Theorem 4.5 implies that the plane $R \times R$ has exactly as many points as the line R, for we have by Theorems 4.2 and 4.5 that $|R \times R| = |R|.|R| = cc = c$.

Let R^3 be the set of all ordered triplets $\langle x, y, z \rangle$ where $x, y, z \in R$. We shall call R^3 the three-dimensional Euclidean space. Any ordered triplet $\langle x, y, z \rangle$ of real numbers will be called a *point* of R^3. The terms x, y, z will be called the *coordinates* of this point.

THEOREM 4.6. *The three-dimensional Euclidean space has exactly as many points as R.*

Proof. Consider the product of the two sets $R \times R$ and R, that is, $(R \times R) \times R$. The elements of this product are pairs, the first term of such a pair is an element of $R \times R$, hence it is itself a pair $\langle x, y \rangle$, and the second term is just a number z in R. Hence every element of $(R \times R) \times R$ is of the form $\langle \langle x, y \rangle, z \rangle$ where $x, y, z \in R$. The mapping

$$f: R^3 \to (R \times R) \times R \quad \text{such that} \quad f(\langle x, y, z \rangle) = \langle \langle x, y \rangle, z \rangle$$

is clearly a bijection. It follows that

$$|R^3| = |(R \times R) \times R| = (|R|.|R|)|R| = (cc)c = cc = c = |R|, \text{ by}$$

Theorems 4.2 and 4.5.

2. Exponentiation

We wish to associate with every two cardinal numbers α, β a cardinal number β^α (β to the power α). The meaning of β^α should be the usual one when $\alpha, \beta \in N$. But if we have two finite sets A, B such that $|A| = k$, $|B| = n$, then, as we have noticed previously (Theorem 2.2, p. 17), there are exactly n^k distinct mappings $f: A \to B$. Equivalently, we can say that there are exactly n^k subsets f of $A \times B$ such that $f: A \to B$. We accept now

DEFINITION 4.2. If A, B are any sets, then we denote by B^A the set defined by

$$\forall f: f \in B^A \Leftrightarrow f: A \to B.$$

In other words, B^A is the set of all graphs of mappings from A to B. We know that $|B^A| = |B|^{|A|}$ when A and B are finite. This observation will serve as basis for

DEFINITION 4.3. Let α, β be cardinal numbers. If A, B are any sets such that $|A| = \alpha$, $|B| = \beta$, then we define

$$\beta^\alpha = |B^A|.$$

We have to show that β^α is well defined—we have to prove

THEOREM 4.7. *If* $|A_1| = |A|$ *and* $|B_1| = |B|$, *then* $|B_1^{A_1}| = |B^A|$.

Proof. Let us show first that $|B^A| \leqslant |B_1^{A_1}|$. Let us define an injection $\mathfrak{F}: B^A \to B_1^{A_1}$. We draw a diagram (Figure 10) on

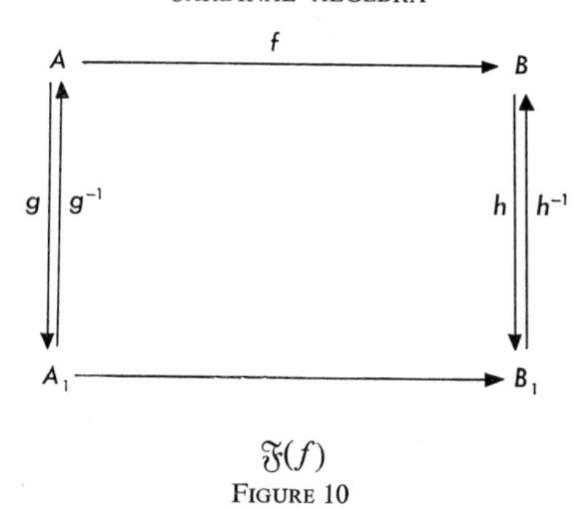

$\mathfrak{F}(f)$
FIGURE 10

which $g : A \to A_1$ and $h : B \to B_1$ are bijections. For any $f : A \to B$, we have to define the image $\mathfrak{F}(f) : A_1 \to B_1$. Now let $f : A \to B$ be given. If $x \in A_1$, then we can assign to x first the element $g^{-1}(x)$ in A, then $f(g^{-1}(x))$ in B and finally $h(f(g^{-1}(x))$ in B_1. In this way we obtain the mapping $hfg^{-1} : A_1 \to B_1$. We define now $\mathfrak{F}(f) = hfg^{-1}$ for every $f : A \to B$. It remains to prove that $\mathfrak{F} : B^A \to B_1^{A_1}$ is injective. For this purpose let us take any two distinct mappings $f_1 : A \to B$, $f_2 : A \to B$. Then there exists an a in A such that $f_1(a) \neq f_2(a)$. Since $g^{-1} : A_1 \to A$ is surjective, we have $a = g^{-1}(a_1)$ for some $a_1 \in A_1$, whence $f_1(g^{-1}(a_1)) \neq f_2(g^{-1}(a_1))$.

Since $h : B \to B_1$ is injective, it follows now that

$$h(f_1(g^{-1}(a_1))) \neq h(f_2(g^{-1}(a_1))),$$

that is

$$(hf_1g^{-1})(a_1) \neq (hf_2g^{-1})(a_1),$$

whence $\mathfrak{F}(f_1) \neq \mathfrak{F}(f_2)$. Thus $\mathfrak{F} : B^A \to B_1^{A_1}$ is injective.

Our assumptions about A_1, B_1 and A, B are symmetric (they will remain unchanged if we simultaneously interchange A_1 with A and B_1 with B). Hence we must have also $|B_1^{A_1}| \leqslant |B^A|$. This inequality and the one proved above yield $|B_1^{A_1}| = |B^A|$.

THEOREM 4.8. *If A, B are any sets, then* $|B|^{|A|} = |B^A|$.

Proof. Denote $|A|$ and $|B|$ by α and β respectively and use Definition 4.3.

Now consider the special case when $A = N$ and $B = \{0, 1\}$. Then we have $|\{0, 1\}^N| = 2^{|N|} = 2^{\aleph_0}$. We shall use these equalities in order to prove

THEOREM 4.9. $2^{\aleph_0} = c$.

Proof. We have to show that $|\{0, 1\}^N| = |I|$. We prove first that $|\{0, 1\}^N| \leqslant |I|$, that is, we define an injection $H: \{0, 1\}^N \to I$. Suppose that $f \in \{0, 1\}^N$, so that $f: N \to \{0, 1\}$. Then we assign to f the real number $H(f)$ whose decimal expansion is

$$H(f) = 0{\cdot}(f(0)+1)(f(1)+1)(f(2)+1)\ldots(f(n)+1)\ldots.$$

This is a proper expansion since each digit is either 1 or 2. Moreover $H(f) \in \left[\frac{1}{9}, \frac{2}{9}\right] \subset I$. We denote the mapping which assigns $H(f)$ to f by $H: \{0, 1\}^N \to I$. If we take two distinct $f_1, f_2 \in \{0, 1\}^N$, then $f_1(n) \neq f_2(n)$ for some n. Consequently $H(f_1)$ and $H(f_2)$ have distinct proper decimal expansions and thus $H(f_1) \neq H(f_2)$. We have shown that $H: \{0, 1\}^N \to I$ is injective.

Now let us prove that $|I| \leqslant |\{0, 1\}^N|$. We shall define an injection $F: I \to \{0, 1\}^N$; the image by this mapping of a number $x \in I$ will be denoted by F_x (instead of the usual $F(x)$). To define F_x consider the binary expansion of x

$$x = 0.a_0a_1a_2\ldots a_n\ldots \qquad (a_n = 0 \text{ or } 1).$$

In other words, we write

$$x = \sum_{n=0}^{\infty} 2^{-n-1}a_n = \frac{a_0}{2} + \frac{a_1}{2^2} + \frac{a_2}{2^3} + \ldots + \frac{a_n}{2^{n+1}} + \ldots$$

where each $a_n = 0$ or 1. For example on Figure 11 we have

$$x = \frac{1}{2} + \frac{0}{4} + \frac{0}{8} + \frac{1}{16} + \frac{1}{32} + \ldots$$

whence the binary expansion of x begins with $x = 0.10011\ldots$.

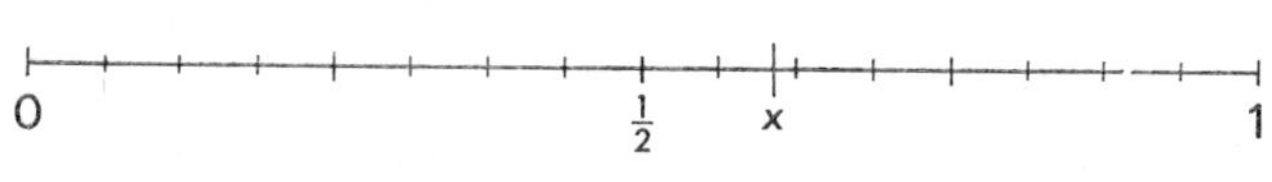

FIGURE 11

We call the expansion of x *proper* if $a_n \neq 1$ holds for infinitely many n. Then every x in I has exactly one proper binary expansion. We assume that the above expansion of x is proper. We assign now to x the element $F_x \in \{0, 1\}^N$ such that $F_x : N \to \{0, 1\}$ is the mapping given by $F_x(n) = a_n$, for every n. If $x \neq x'$ and x' has the proper binary expansion

$$x' = 0.a'_0a'_1a'_2\ldots a'_n\ldots$$

then $a_n \neq a'_n$ for some n. Thus $F_x(n) \neq F_{x'}(n)$, which shows that $F_x \neq F_{x'}$. We have proved that $F: I \to \{0, 1\}^N$ is injective. Applying the CBS-theorem, we see that $|I| = |\{0, 1\}^N|$.

3. The set of all subsets of A

For any given set A we shall consider the set of all subsets of A. We denote the latter by $\mathcal{B}(A)$, in honour of G. Boole (1813-64), who systematically studied the properties of this set.

DEFINITION 4.4. $\forall B : B \in \mathcal{B}(A) \leftrightarrow B \subset A$.

We note that $\varnothing \in \mathcal{B}(A)$. If $A = \{1, 2, 3\}$, then

$\mathcal{B}(\{1, 2, 3\}) = \{(\varnothing, \{1\}, \{2\}, \{3\}, \{1, 2\}, \{1, 3\}, \{2, 3\}, \{1, 2, 3\}\}$

whence $|\mathcal{B}(\{1, 2, 3\})| = 8$.

This equality is a special case of

THEOREM 4.10. $|\mathcal{B}(A)| = 2^{|A|}$.

Proof. It is enough to prove that $|\mathcal{B}(A)| = |\{0, 1\}^A|$. Let us assign to every $B \subset A$ the function on A that is equal to 1 on B and equal to 0 on $A-B$. This function is called the *characteristic function of B* and is usually denoted by $\chi_B : A \to \{0, 1\}$ so that

$$\chi_B(x) = \begin{cases} 1 & \text{if } x \in B, \\ 0 & \text{if } x \in A-B. \end{cases}$$

The mapping which assigns to every $B \subset A$ the graph χ_B of the characteristic function of B will be denoted by $\chi : \mathcal{B}(A) \to \{0, 1\}^A$. This mapping is injective, for it is clear that different sets have different characteristic functions. The mapping is also surjective, for if we have an arbitrary $f: A \to \{0, 1\}$, then this is the characteristic function of the set

$$B = \{x \in A \mid f(x) = 1\}.$$

THEOREM 4.11. *The set of all subsets of N has cardinality c.*

Proof. From Theorems 4.9 and 4.10 we have

$$|\mathcal{B}(N)| = 2^{|N|} = 2^{\aleph_0} = c.$$

The equality $|\mathcal{B}(N)| = c$, together with $\aleph_0 < c$ imply that $|N| < |\mathcal{B}(N)|$. This inequality is a special case of the following theorem due to Cantor

THEOREM 4.12. $|A| < |\mathcal{B}(A)|$ *for every set A.*

Proof. We show first that $|A| \leqslant |\mathcal{B}(A)|$. Let $g : A \to \mathcal{B}(A)$ be the mapping that assigns to every x in A the one-element subset $\{x\} \subset A$, that is, $g(x) = \{x\}$. Then $g : A \to \mathcal{B}(A)$ is injective, whence $|A| \leqslant |\mathcal{B}(A)|$.

It remains to prove that $|A| \neq |\mathcal{B}(A)|$. We shall assume for the moment that $|A| = |\mathcal{B}(A)|$ and we shall obtain a contradiction. We assume therefore that there exists a bijection $f : A \to \mathcal{B}(A)$. For any x in A, $f(x)$ is a subset of A; let us denote $f(x)$ by B_x. Let us consider any element x in A and its image B_x: Either $x \in B_x$ or $x \notin B_x$. The set of all those x that have the latter property will be denoted by C, so that

$$C = \{x \in A \mid x \notin B_x\}.$$

C is a certain subset of A, and since $f : A \to \mathcal{B}(A)$ is surjective, C must be an image—that is, there exists an a in A such that $C = B_a$. Let us ask now: Does a belong to C? Certainly $a \in C \Leftrightarrow a \notin B_a$, since C is the set of all x satisfying $x \notin B_x$. On the other hand, we have $B_a = C$ so that $a \notin B_a \Leftrightarrow a \notin C$.

We have deduced therefore

$$a \in C \Leftrightarrow a \notin B_a \quad \text{and} \quad a \notin B_a \Leftrightarrow a \notin C,$$

whence $a \in C \Leftrightarrow a \notin C$. This is certainly a contradiction, and therefore $|A| \neq |\mathcal{B}(A)|$.

THEOREM 4.13. $\alpha < 2^\alpha$ *for every cardinal number* α.

Proof. Take a set A s.t. $|A| = \alpha$. Then, by Theorems 4.10 and 4.12

$$\alpha = |A| < |\mathcal{B}(A)| = 2^\alpha.$$

From this theorem we deduce that

$$\aleph_0 < 2^{\aleph_0} \quad (= c) \quad < 2^{(2^{\aleph_0})} < 2^{\left(2^{(2^{\aleph_0})}\right)} < \ldots$$

Here we have written four distinct infinite cardinal numbers. Are these the first four? In particular, does there exist a cardinal number α between $\aleph_0$ and $2^{\aleph_0}$? The statement that no such α exists is called

THE CONTINUUM HYPOTHESIS. *There does not exist a cardinal number* α *such that*

$$\aleph_0 < \alpha < 2^{\aleph_0}.$$

This statement cannot be proved or disproved. In fact it was proved by K. Gödel in 1938 that by accepting the continuum hypothesis as a true statement one does not introduce a contradiction. In 1964 P. J. Cohen proved that no contradiction is introduced if the negation of the continuum hypothesis is accepted as a true statement.

The sets which most mathematicians use in their work are of cardinality not exceeding 2^c, and most of them turn out to be either of cardinality $\aleph_0$ or c.

4. Sequences

Let A be a non-empty set. Intuitively we mean by a sequence of elements of A an infinite row $a_0, a_1, a_2, \ldots, a_n, \ldots$ where each a_n belongs to A. If such a sequence is given then we can assign to it a mapping $f: N \to A$, by putting $f(n) = a_n$ for every n. Conversely, every mapping $f: N \to A$ defines a sequence of elements of A, namely the sequence $f(0), f(1), f(2), \ldots, f(n), \ldots$. Since a sequence and a mapping correspond to each other in this way, there is no need to distinguish between them. We accept therefore

DEFINITION 4.5. Every mapping $f: N \to A$ will be called a *sequence of elements of* A and $f: N \to A$ will be also denoted by $a_0, a_1, a_2, \ldots, a_n, \ldots$ where $a_n = f(n)$ for every n in N.

Since the set of all mappings from N to A has the same cardinal number as the set A^N of all graphs of these mappings, we derive from Definition 4.5 that the set of all sequences of elements of A has cardinality $|A^N|$. In particular, the cardinal number of the set of all sequences of real numbers is $|R^N| = c^{\aleph_0}$ and the cardinal number of the set of all sequences of natural numbers is $|N^N| = \aleph_0^{\aleph_0}$. Let us calculate these.

THEOREM 4.14. $\aleph_0^{\aleph_0} = 2^{\aleph_0}$.

Proof. It is clear that $\{0, 1\}^N \subset N^N \Rightarrow 2^{\aleph_0} \leqslant \aleph_0^{\aleph_0}$. To see that $\aleph_0^{\aleph_0} \leqslant 2^{\aleph_0}$ we observe that $N^N \subset \mathcal{B}(N \times N)$ ($f \in N^N$ means that

f is a graph of a mapping with range N and domain N, whence $f \subset N \times N$). Thus, by Theorem 4. 10 p. 56

$$\aleph_0^{\aleph_0} = |N^N| \leqslant |\mathcal{B}(N\times N)| = 2^{|N\times N|} = 2^{\aleph_0\aleph_0} = 2^{\aleph_0}.$$

THEOREM 4.15. $c^{\aleph_0} = 2^{\aleph_0}$.

Proof. It is clear that $\{0, 1\}^N \subset R^N \Rightarrow 2^{\aleph_0} \leqslant c^{\aleph_0}$. To prove the opposite inequality, we shall define an injection $\mathfrak{F} : R^N \to N^{N\times N}$. Indeed, this will imply that $c^{\aleph_0} \leqslant \aleph_0^{\aleph_0\aleph_0} = \aleph_0^{\aleph_0}$ which is $2^{\aleph_0}$ by the previous theorem. Let f be an arbitrary element of R^N, so that $f: N \to R$. We consider the numbers

$$f(0) = a_0^{(0)}\cdot a_1^{(0)}a_2^{(0)}\ldots a_n^{(0)}\ldots$$
$$f(1) = a_0^{(1)}\cdot a_1^{(1)}a_2^{(1)}\ldots a_n^{(1)}\ldots$$
$$\cdots\cdots$$
$$f(k) = a_0^{(k)}\cdot a_1^{(k)}a_2^{(k)}\ldots a_n^{(k)}\ldots$$
$$\cdots\cdots$$

in their proper decimal expansions so that $a_n^{(k)} \in N$ for every $\langle k, n\rangle \in N\times N$ and $a_n^{(k)} \leqslant 9$ if $n > 0$. We assign now to f the element $\mathfrak{F}_f \in N^{N\times N}$ such that the function $\mathfrak{F}_f : N\times N \to N$ is defined by

$$\mathfrak{F}_f\langle k, n\rangle = a_n^{(k)} \quad \text{for all} \quad \langle k, n\rangle \in N\times N.$$

We denote the mapping which assigns $\mathfrak{F}_f$ to f by $\mathfrak{F} : R^N \to N^{N\times N}$. This mapping is injective, for if we take $f_1 \neq f$ in R^N, then the mappings $f_1 : N \to R, f: N \to R$ are different and hence for some $k, f_1(k) \neq f(k)$. If the proper decimal expansion of $f_1(k)$ is

$$f_1(k) = b_0^{(k)}\cdot b_1^{(k)}b_2^{(k)}\ldots b_n^{(k)}\ldots$$

then it follows that $a_n^{(k)} \neq b_n^{(k)}$ for some n. Thus $\mathfrak{F}_{f_1}\langle k, n\rangle \neq \mathfrak{F}_f\langle k, n\rangle$ which shows that $\mathfrak{F}_{f_1} \neq \mathfrak{F}_f$.

EXERCISES

1. Let $B \subset R^3$ be the unit ball and let $\mathring{B}$ be its interior, defined by

$$B = \{\langle x, y, z\rangle \in R^3 \mid x^2+y^2+z^2 \leqslant 1\},$$
$$\mathring{B} = \{\langle x, y, z\rangle \in R^3 \mid x^2+y^2+z^2 < 1\}.$$

Prove that $|B| = |\mathring{B}| = |B-\mathring{B}| = c$. ($B-\mathring{B}$ is the two-dimensional sphere.)

2. Find the cardinal number of the set T of all those points in R^3 all of whose three coordinates are rational. Deduce that if Ω is any set of pairwise disjoint balls in R^3, then $|\Omega| \leqslant \aleph_0$.

3. Let α, β, γ be any cardinal numbers.

(i) Prove that $\alpha\beta = \beta\alpha$.
(ii) Prove that $(\alpha\beta)\gamma = \alpha(\beta\gamma)$.
(iii) Prove that $\alpha \leqslant \beta \Rightarrow \alpha\gamma \leqslant \beta\gamma$.
(iv) Find $\aleph_0 c$, $n\aleph_0$ and nc if $n \in N$.

4. Show that $\alpha^\beta\alpha^\gamma = \alpha^{\beta+\gamma}$ for any cardinal numbers α, β, γ. Deduce that

(i) $\aleph_0^n = \aleph_0$,
(ii) $c^n = c$,

for every n in N.

5. Let $\mathcal{B}_f(A)$ be the set of all finite subsets of A. Prove that

(i) $|\mathcal{B}_f(N)| = \aleph_0$,
(ii) $|\mathcal{B}_f(R)| = c$.

6. Find the cardinal number of the set of all

(i) mappings $f: Q \to R$,
(ii) injections $f: N \to N$,
(iii) continuous functions $f: R \to R$,
(iv) continuous bijections $f: R \to R$.

About Wrong Definitions

We have described on previous pages various methods of obtaining sets. We assumed that certain sets, like N, Z, Q, R exist and have certain properties and we introduced methods of forming new sets from any given ones: by taking unions, intersections, products, subsets or forming the set of all subsets of a given set. The purpose of this section is to give a warning that another method (not one of those described above) of defining a set may lead to contradictions.

A WRONG DEFINITION: Let $p(x)$ denote a meaningful statement about x. That is, if we replace x by the name of some object, then this statement will become either a true or a false proposition, depending on the object chosen. Then a set S is defined by the following condition

$$\forall x : x \in S \Leftrightarrow p(x) \quad \text{is true.}$$

We write $S = \{x \mid p(x)\}$.

We have used in this book definitions of the above type; the union, product and the set of all subsets where defined as

$$A \cup B = \{x \mid x \in A \quad \text{or} \quad x \in B\}$$

$$A \times B = \{x \mid \exists a \in A \quad \text{and} \quad \exists b \in B \quad \text{s.t.} \quad x = \langle a, b \rangle\}$$

$$\mathcal{B}(A) = \{x \mid x \subset A\}.$$

But apart from the special cases which appeared in this book, definitions of the above type should be avoided, as they may lead to contradictions. Those contradictions are known as the antinomies. We give two examples.

The antinomy of the set of all sets. Let S be the set of all sets, that is,

$$S = \{x \mid x \text{ is a set}\}.$$

Consider the set $\mathcal{B}(S)$ of all subsets of S. Clearly

$$\forall x: \quad x \in \mathcal{B}(S) \Rightarrow x \text{ is a set} \Rightarrow x \in S.$$

Therefore $\mathcal{B}(S) \subset S$ and hence $|\mathcal{B}(S)| \leqslant |S|$. But this contradicts Theorem 4.12 on p. 57.

The antinomy of Bertrand Russell. Let S be the set of those sets which are not elements of themselves, that is

$$S = \{x \mid x \text{ is a set and } x \notin x\}.$$

The condition that $x \notin x$ is satisfied by all sets which we have considered—for example, $N \notin N$, $R \notin R$. Let us ask: Is S an element of S? We have

$$S \in S \Leftrightarrow S \text{ is a set and } S \notin S$$

and since S is a set, by definition, we finally get

$$S \in S \Leftrightarrow S \notin S.$$

This is a contradiction.

Remark. If A is a set and $p(x)$ is a meaningful statement about x, then it is correct to define a set S by

$$\forall x: \quad x \in S \Leftrightarrow x \in A \text{ and } p(x) \text{ is true}.$$

For this set we have accepted the notation $S = \{x \in A \mid p(x)\}$. No contradiction is known to result from such a definition.

Hints and Answers

1. Sets

1. $\{1, 2\}, \{2, 3\}, \{1, 2, 3\} = \{1, 2\} \cup \{2, 3\}$

$$\{1\} = \{1, 2\} - \{2, 3\}$$
$$\{2\} = \{1, 2\} \cap \{2, 3\}$$
$$\{3\} = \{2, 3\} - \{1, 2\}$$
$$\{1, 3\} = \{1\} \cup \{3\} = (\{1, 2\} - \{2, 3\}) \cup (\{2, 3\} - \{1, 2\})$$
$$\varnothing = \{1\} \cap \{3\} = (\{1, 2\} - \{2, 3\}) \cap (\{2, 3\} - \{1, 2\}).$$

2. $A_0 = \{0, 1, 3, 5, 7, \ldots\}$

$A_1 = \{2, 2\times 3, 2\times 5, 2\times 7, \ldots\}$

$A_k = \{2^k, 2^k\times 3, 2^k\times 5, 2^k\times 7, \ldots\}$

3. $|A\cup B| = 13$ (see Exercise 4).
4. $|A\cup B| = |A|+|B|-|A\cap B|$.
5. $|A\cup B\cup C| = |A|+|B|+|C|-|A\cap B|-|B\cap C|- -|C\cap A|+|A\cap B\cap C|$.
6. Let S = set of all combatants; denote $|S| = n$. We have $A, B, C, D \subset S$ such that $|A| = 0.7n, |B| = 0.8n, |C| = 0.75n, |D| = 0.85n$ and $x = \frac{100}{n}|A\cap B\cap C\cap D|$. From Exercise 4,

$$|A \cap B| = |A|+|B|-|A \cup B| \geqslant 0.7n+0.8n-n = 0.5n,$$
$$|A \cap B \cap C| = |(A \cap B) \cap C| = |A \cap B|+|C|$$
$$-|(A \cap B) \cup C| \geqslant 0.5n+0.75n-n = 0.25n,$$
$$|A \cap B \cap C \cap D| = |A \cap B \cap C|+|D|-|(A \cap B \cap C) \cup D|$$
$$\leqslant 0.25n+0.85n-n = 0.1n.$$

Hence $x \geqslant 10$ and this is the minimal value since it is possible to have everywhere above $=$ instead of $\geqslant$.

7. Let S = set of all students. We have A, B, $C \subset S$ s.t. $|A| = 28$, $|B| = 30$, $|C| = 42$, $|A \cap B| = 8$, $|A \cap C| = 10$, $|B \cap C| = 5$, $|A \cap B \cap C| = 3$.

(i) $A \cup B \cup C$ = set of those who study some language. From Exercise 5, $|A \cup B \cup C| = 28+30+42-8-10-5+3 = 80$.

Thus 20 were studying no language.

(ii) $C-(A \cup B)$ = set of those studying French only. Apply to the three sets $C \cap A$, $C \cap B$, $C-(A \cup B)$ whose union is C the formula in Exercise 5,

$$|C| = |C \cap A|+|C \cap B|+|C-(A \cup B)|-|A \cap B \cap C|-$$
$$|\varnothing|-|\varnothing|+|\varnothing|,$$

that is, $42 = 10+5+|C-(A \cup B)|-3$, whence $|C-(A \cup B)| = 30$.

8. (i) $x \in A \cap (B \cup C) \Leftrightarrow x \in A$ and $x \in B \cup C \Leftrightarrow x \in A$ and $(x \in B$ or $x \in C) \Leftrightarrow (x \in A$ and $x \in B)$ or $(x \in A$ and $x \in C)$ $\Leftrightarrow x \in A \cap B$ or $x \in A \cap C \Leftrightarrow x \in (A \cap B) \cup (A \cap C)$.

(ii) $x \in A \cup (B \cap C) \Leftrightarrow x \in A$ or $(x \in B$ and $x \in C)$ $\Leftrightarrow (x \in A$ or $x \in B)$ and $(x \in A$ or $x \in C) \Leftrightarrow x \in A \cup B$ and $x \in A \cup C \Leftrightarrow x \in (A \cup B) \cap (A \cup C)$.

9. $A \cup B = B \Leftrightarrow A \cap B = A \cup A - B = \varnothing \Leftrightarrow A \subset B.$
$A \cup B = A \cap B \Leftrightarrow (A-B) \cup (B-A) = \varnothing \Leftrightarrow A = B.$

10. $A_k = (0, 1/(k+1)).$

2. Products and Mappings

1. $(h(gf))(x) = h((gf)(x)) = h(g(f(x))),$
$((hg)f)(x) = (hg)(f(x)) = h(g(f(x))),$

whence $(h(gf))(x) = ((hg)f)(x)$ for $\forall x \in A$. Thus $h(gf) = (hg)f$.

2. $(fg)(1) = f(g(1)) = f(2) = 3, \quad (fg)(2) = f(g(2)) = f(1) = 2,$
$(fg)(3) = f(g(3)) = f(3) = 1.$ Hence
$$fg = \{\langle 1, 3\rangle \langle 2, 2\rangle, \langle 3, 1\rangle\}.$$

Similarly

$$gf = \{\langle 1, 1\rangle, \langle 2, 3\rangle, \langle 3, 2\rangle\}$$
$$ff = \{\langle 1, 3\rangle, \langle 2, 1\rangle, \langle 3, 2\rangle\}$$
$$fgf = f(gf) = \{\langle 1, 2\rangle, \langle 2, 1\rangle, \langle 3, 3\rangle\} = g$$

and, since $f(1) = 2 \Leftrightarrow 1 = f^{-1}(2)$, $f(2) = 3 \Leftrightarrow 2 = f^{-1}(3)$, etc.
$$f^{-1} = \{\langle 2, 1\rangle, \langle 3, 2\rangle, \langle 1, 3\rangle\} = ff.$$

3. (i) Let $y \in C$. $g : B \to C$ surjective $\Rightarrow \exists z \in B$ s.t. $y = g(z)$. Since $f : A \to B$ surjective, $\exists x \in A$ s.t. $z = f(x)$. Thus $y = g(f(x)) = (gf)(x)$.

(ii) Let $x, y \in A$. Then $x \neq y \Rightarrow f(x) \neq f(y) \Rightarrow g(f(x)) \neq g(f(y)) \Rightarrow (gf)(x) \neq (gf)(y)$, whence $gf : A \to C$ is injective.

4. Let $g, h : N \to N$, $g \neq h$. Then $\exists x \in N$ s.t. $g(x) \neq h(x)$. Since $f : N \to N$ is injective, we must have $f(g(x)) \neq f(h(x))$ whence $fg \neq fh$. An example of $gf = hf$ is given by $g(x) = h(x) = x$ if $x \neq 2$ and $g(2) \neq h(2)$. Since 2 is not a square of an integer, we have $g(x^2) = h(x^2) = x^2$, whence $(gf)(x) = g(f(x)) = g(x^2) = h(x^2) = h(f(x)) = (hf)(x)$.

5. Let $g, h : N \to N$, $g \neq h$. Then $\exists y \in N$ s.t. $g(y) \neq h(y)$. Since $f : N \to N$ is surjective, $\exists x \in N$ s.t. $f(x) = y$. Thus $g(f(x)) \neq h(f(x))$, whence $gf \neq hf$. If $g(x) = 2x+1$, $h(x) = 2(2x+1)$, then $fg = fh$.

6. $gf : A \to C$ is bijective by Exercise 3. Using Theorem 2.3 p. 20 several times, we get $y = (gf)^{-1}(x) \Leftrightarrow x = (gf)(y) \Leftrightarrow x = g(f(y)) \Leftrightarrow f(y) = g^{-1}(x) \Leftrightarrow y = f^{-1}(g^{-1}(x)) \Leftrightarrow y = (f^{-1}g^{-1})(x)$, whence $(gf)^{-1} = g^{-1}f^{-1}$.

7. (i) Exercise 4, (ii) Exercise 5.

8. $f(n) = n+k$ for every n.

9. $f(n) = 2n$ and $g(n) = 2n+1$ for every n.

10. Suppose $f(A_1) \cap f(A_2) \neq \varnothing$. Then $\exists b \in f(A_1) \cap f(A_2)$, that is, $\exists a_1 \in A_1$, $a_2 \in A_2$ s.t. $b = f(a_1) = f(a_2)$. But $a_1 \neq a_2$ since $A_1 \cap A_2 = \varnothing$. This contradicts the injectivity of $f : A \to B$. Thus $f(A_1) \cap f(A_2) = \varnothing$. Clearly $A_1, A_2 \subset A \Rightarrow f(A_1) \cup f(A_2) \subset f(A)$. To prove $f(A) \subset f(A_1) \cup f(A_2)$, assume $b \in f(A)$. Then $\exists a \in A$ s.t. $b = f(a)$. We have $a \in A_1$ or $a \in A_2$, whence $b = f(a) \in f(A_1)$ or $f(A_2) \Rightarrow b \in f(A_1) \cup f(A_2)$. Both inclusions imply equality.

11. Let $x_1 \neq x_2$. The function is increasing. Hence $g(x_1) > g(x_2)$ if $x_1 > x_2$ and $g(x_1) < g(x_2)$ if $x_1 < x_2$. In either case $g(x_1) \neq g(x_2)$, hence the function is injective.

Let $y \in R$, we should show that y is an image. If $y \geqslant 0$, then $y = g(x)$ for $x = y/(1+y)$, by

$$g\left(\frac{y}{1+y}\right) = \frac{y/(1+y)}{1-y/(1+y)} = y.$$

If $y < 0$, then $y = g(x)$ for $x = y/(1-y)$, by

$$g\left(\frac{y}{1-y}\right) = \frac{y/(1-y)}{1-|y/(1-y)|} = \frac{y/(1-y)}{1+y/(1-y)} = y.$$

Hence the function is surjective.

3. Cardinal Numbers

1. $A \cup B = A \cup (B-A)$ and $A \cap (B-A) = \varnothing \Rightarrow |A \cup B| = |A|+|B-A| \leqslant |A|+|B| = \aleph_0 + \aleph_0 = \aleph_0$ (using Theorem 3.18, p. 39 and Theorem 3.12, p. 36). Also $A \subset A \cup B \Rightarrow |A| \leqslant |A \cup B| \Rightarrow \aleph_0 \leqslant |A \cup B|$. By the CBS-theorem (Theorem 3.24, p. 43) it follows now that $|A \cup B| = \aleph_0$.

2. $S \subset R \Rightarrow |S| \leqslant |R| \Rightarrow |S| \leqslant c$.

$(a, b) \subset S \Rightarrow |(a, b)| \leqslant |S| \Rightarrow c \leqslant |S|$. CBS-theorem $\Rightarrow |S| = c$. To prove $c+c = c$, take $S = (a, b) \cup (c, d)$ where $a < b < c < d$. By the above $c = |S| = |(a, b)|+|(c, d)| = c+c$.

3. $\sqrt{2}$ is irrational $\Rightarrow \sqrt{2} \in R-Q$. Thus $n\sqrt{2} \in R-Q$ for all $n \in N_1$ (see Exercise 8, p. 24). But $|\{n\sqrt{2} \in R|\ n \in N_1\}| = |N_1| = \aleph_0$ whence $\aleph_0 \leqslant |R-Q|$. From Theorem 3.25, p. 45 and Theorem 3.19, p. 40 follows now

$$|R-Q|+|Q| = |R-Q|+\aleph_0 = |R-Q|.$$

Also $|R-Q|+|Q| = |(R-Q) \cup Q| = |R| = c$, whence $|R-Q| = c$.

4. $|A|+|B| = |A|+|(B-A)\cup(A\cap B)| = |A|+|B-A| + |A\cap B| = |A\cup(B-A)|+|A\cap B| = |A\cup B|+|A\cap B|$.

5. $|A\cup B| \leqslant |A|+|B|$ as in Exercise 1 above. Using Theorems 3.19 or 3.20 (p. 40), $|A|+|B| = |A|$. Hence $|A\cup B| \leqslant |A|$. Trivially $|A| \leqslant |A\cup B|$, whence equality.

6. $\alpha \leqslant \beta \Rightarrow \exists\delta$ s.t. $\alpha+\delta = \beta$ (Theorem 3.16, p. 37). Hence $\gamma+\alpha = \alpha \Rightarrow \gamma+\beta = \gamma+(\alpha+\delta) = (\gamma+\alpha)+\delta = \alpha+\delta = \beta$.

4. Cardinal Algebra

1. $B, \mathring{B}, B-\mathring{B} \subset R^3 \Rightarrow |B|, |\mathring{B}|, |B-\mathring{B}| \leqslant c$. The mapping $g:(0, 1) \to B$ defined by $g(x) = \langle x, 0, 0\rangle$ is injective $\Rightarrow c \leqslant |\mathring{B}| \leqslant |B|$, whence $|B| = |\mathring{B}| = c$. The mapping $f:(0,\pi) \to B-\mathring{B}$ defined by $f(x) = \langle \sin x, \cos x, 0\rangle$ is injective $\Rightarrow c \leqslant |B-\mathring{B}|$, whence $|B-\mathring{B}| = c$.

2. $T = \{\langle x, y, z\rangle \in R^3 \mid x, y, z \in Q\}$. The mapping $h: T \to (Q\times Q)\times Q$ given by $h(\langle x, y, z\rangle) = \langle\langle x, y\rangle, z\rangle$ is bijective. Hence $|T| = |Q\times Q|.|Q| = (\aleph_0\aleph_0)\aleph_0 = \aleph_0\aleph_0 = \aleph_0$. Let $f: T \to N$ be bijection. If B is a ball belonging to Ω, then $B\cap T \neq \varnothing$. Let $g:\Omega \to T$ be the function which assigns to $B\in\Omega$ this point $g(B) = \langle x, y, z\rangle \in B\cap T$ for which $f(\langle x, y, z\rangle) \in N$ is smallest. Then $g:\Omega \to T$ is injective, for if $B_1 \neq B_2$ are in Ω, then $B_1\cap B_2 = \varnothing \Rightarrow g(B_1) \neq g(B_2)$.

3. (i) $\exists$ bijection $h: A\times B \to B\times A$ defined by $h(\langle x, y\rangle) = \langle y, x\rangle$.

(ii) $\exists$ bijection $h: A\times(B\times C) \to (A\times B)\times C$ defined by $h(\langle x, \langle y, z\rangle\rangle) = \langle\langle x, y\rangle, z\rangle$.

(iii) Let $|A| = \alpha$, $|B| = \beta$, $|C| = \gamma$ and let $f: A \to B$ be injective. Define a mapping $A \times C \to B \times C$ by $g(\langle x, y\rangle) = \langle f(x), y\rangle$. This mapping is injective.

(iv) From part (iii) follows $c \leqslant cn \leqslant c\aleph_0 \leqslant cc = c$ (Theorem, 4.5, p. 51). Hence $cn = c\aleph_0 = c$, by the CBS-theorem. From part (iii) follows $\aleph_0 \leqslant \aleph_0 n \leqslant \aleph_0\aleph_0 = \aleph_0$ (Theorem 4.3, p. 49). Hence $\aleph_0 n = \aleph_0$, by the CBS-theorem.

4. Let $|A| = \alpha$, $|B| = \beta$ and $|C| = \gamma$, where $B \cap C = \varnothing$ so that $\beta + \gamma = |B \cup C|$. Then

$$\alpha^\beta \alpha^\gamma = |A^B|\,|A^C| = |A^B \times A^C|, \qquad \alpha^{\beta+\gamma} = |A^{B\cup C}|.$$

A bijection $F: A^B \times A^C \to A^{B\cup C}$ is defined as follows: If $\langle f_1, f_2\rangle \in A^B \times A^C$, that is, if $f_1: B \to A$ and $f_2: C \to A$, then $F(\langle f_1, f_2\rangle): B \cup C \to A$ is the mapping which is equal to f_1 on B and equal to f_2 on C,

$$F(\langle f_1, f_2\rangle)(x) = \begin{cases} f_1(x) \text{ if } x \in B, \\ f_2(x) \text{ if } x \in C. \end{cases}$$

(i) Induction on n. $\aleph_0^1 = |N^{\{0\}}|$. $\exists$ bijection $N^{\{0\}} \to N$ which assigns to every $f: \{0\} \to N$ the number $f(0)$. Hence $\aleph_0^1 = \aleph_0$. Assuming theorem true for n, that is $\aleph_0^n = \aleph_0$, we have, by the first part of this question

$$\aleph_0^{n+1} = \aleph_0^n \aleph_0^1 = \aleph_0^n \aleph_0 = \aleph_0 \aleph_0 = \aleph_0.$$

(ii) Analogous to (i).

5. (i) Let $\mathcal{B}_k(N) = \{A \in \mathcal{B}(N) \mid |A| = k\}$ and $I_k = \{\{n+1, n+2, \ldots, n+k\} \mid n \in N\} \subset \mathcal{B}_k(N)$ for every $k \in N$, $k > 0$. Clearly $|I_k| = \aleph_0$ whence $\aleph_0 \leqslant |\mathcal{B}_k(N)|$. To prove the opposite inequality write every k-element subset ($k > 0$) of N in increasing order as

$x_0 < x_1 < \ldots < x_{k-1}$ and assign to $\{x_0, x_1, \ldots, x_{k-1}\}$ the mapping $f\colon \{0, 1, \ldots, k-1\} \to N$ defined by $f(0) = x_0$, $f(1) = x_1$, $\ldots$, $f(k-1) = x_{k-1}$. This gives an injection $\mathcal{B}_k(N) \to N^{\{0,1,\ldots,k-1\}}$ whence $|\mathcal{B}_k(N)| \leqslant \aleph_0^k = \aleph_0$ (by Exercise 4(i)). Thus $|\mathcal{B}_k(N)| = \aleph_0$ for every k. Consider now the sets $A_0, A_1, A_2, \ldots$ defined on p. 30 (Figure 7). Since $|A_k| = \aleph_0$, there exist bijections $A_k \to \mathcal{B}_{k+1}(N)$ for every k. Since all the sets A_k are pairwise disjoint, and so are the sets $\mathcal{B}_k(N)$, these bijections define a bijection

$$N \times N = \bigcup_{k=0}^{\infty} A_k \to \bigcup_{k=1}^{\infty} \mathcal{B}_k(N).$$

Thus $\left|\bigcup_{k=0}^{\infty} \mathcal{B}_k(N)\right| = |N \times N| = \aleph_0$ whence $|\mathcal{B}_f(N)| = \aleph_0 + |\mathcal{B}_0(N)| = \aleph_0 + |\{\varnothing\}| = \aleph_0 + 1 = \aleph_0$.

(ii) Let $\mathcal{B}_k(R) = \{A \in \mathcal{B}(R) \mid |A| = k\}$ and $I_k = \{\{a+1, a+2, \ldots, a+k\} \mid a \in R\} \subset \mathcal{B}_k(R)$. Clearly $|I_k| = c$ whence $c \leqslant |\mathcal{B}_k(R)| \leqslant \left|\bigcup_{k=1}^{\infty} \mathcal{B}_k(R)\right|$. As in part (i) define an injection $\mathcal{B}_k(R) \to R^{\{0,1,\ldots,k-1\}}$ whence $|\mathcal{B}_k(R)| \leqslant c^k = c$ (by Exercise 4(ii)). Thus $|\mathcal{B}_k(R)| = c$. Let $B_k = \{\langle x, k\rangle \in R \times R \mid x \in R\}$ (= a line parallel to the x-axis). There exist bijections $\mathcal{B}_k(R) \to B_k$ and since the sets B_k are pairwise disjoint, and so are the sets $\mathcal{B}_k(R)$, these bijections define a bijection

$$\mathcal{B}_f(R) = \bigcup_{k=1}^{\infty} \mathcal{B}_k(R) \to \bigcup_{k=1}^{\infty} B_k \subset R \times R$$

whence $\left|\bigcup_{k=1}^{\infty} \mathcal{B}_k(R)\right| \leqslant c$. Thus $\bigcup_{k=1}^{\infty} \mathcal{B}_k(R) = c$ and $|\mathcal{B}_f(R)| = c + |\mathcal{B}_0(R)| = c+1 = c$.

6. (i) The cardinal number is that of R^Q, which is $c^{\aleph_0}$. From Theorem 4.15, p. 60 and Theorem 4.9, p. 54 we have $c^{\aleph_0} = 2^{\aleph_0} = c$.

(ii) Let $I(N) = \{f \in N^N \mid f: N \to N \text{ is injective}\}$. Clearly $I(N) \subset N^N \Rightarrow |I(N)| \leqslant |N^N|$. Define now $\mathfrak{F}: N^N \to I(N)$ to be the mapping which asigns to $f \in N^N$ the image $\mathfrak{F}_f \in I(N)$ defined by

$$\mathfrak{F}_f(n) = f(0)+f(1)+ \cdots +f(n)+n.$$

Then $\mathfrak{F}_f \in N^N$ and $\mathfrak{F}_f(n+1)-\mathfrak{F}_f(n) = f(n+1)+1$, whence $\mathfrak{F}_f$ is increasing, and therefore injective. To prove that $\mathfrak{F}: N^N \to I(N)$ is injective, take $f_1 \neq f_2$ in N^N. Take the smallest number $n \in N$ s.t. $f_1(n) \neq f_2(n)$. Then $\mathfrak{F}_{f_1}(n) \neq \mathfrak{F}_{f_2}(n)$ by the above formula. Hence $f_1 \neq f_2 \Rightarrow \mathfrak{F}_{f_1} \neq \mathfrak{F}_{f_2}$. $\exists$ injective $\mathfrak{F}: N^N \to I(N) \Rightarrow |N^N| \leqslant |I(N)|$. Thus $|I(N)| = |N^N| = \aleph_0^{\aleph_0} = c$ (see Theorem 4.14, p. 59).

(iii) Let $C = \{f \in R^R \mid f: R \to R \text{ is continuous}\}$. Assign to each $f \in C$ the function $f_0: Q \to R$ such that $f_0(x) = f(x)$ for all x in Q. This defines a mapping $C \to R^Q$. This mapping is injective, for if $f \neq g$ are in C, then $\exists\, x_0 \in R$ s.t. $f(x_0) \neq g(x_0)$, say $f(x_0) < g(x_0)$ and then, by continuity, $\exists \delta > 0$ s.t. $f(x) < g(x)$ for all $x \in (x_0-\delta,\ x_0+\delta)$. Picking a rational $x \in (x_0-\delta,\ x_0+\delta)$ we get $f_0(x) < g_0(x)$, whence $f_0 \neq g_0$. Hence $|C| \leqslant |R^Q| = c^{\aleph_0} = c$ (see Theorem 4.15, p. 60). Let $D \subset C$ be the subset of constant functions. Since $|D| = c$, we have $c \leqslant |C|$, whence $|C| = c$.

(iv) Let $B = \{f \in R^R \mid f: R \to R \text{ is continuous and bijective}\}$. Then $B \subset C$ implies $|B| \leqslant c$. Consider for every $a \in R$ the function $f: R \to R$ such that $f(x) = x+a$. The set of these (linear) functions is of cardinal number c and it is contained in B. Hence $c \leqslant |B|$.

Index

Aleph zero 26

Bijection 20

Characteristic function of a set 56
Complementary subsets 24
Composite mapping 22
Continuum 26
Continuum hypothesis 58

Difference 10
Disjoint sets 8
Domain of a mapping 17

Empty set 7
Euclidean plane 14
Euclidean three-dimensional space 51
Exponentation 52

Finite sets 39
Function 15

Graph of a function 17

Identity mapping 17
Image 16
Inclusion 4
Inequality 32
Infinite sets 39
Injection 19
Intersection 5
Interval 3
Inverse mapping 20

Mapping 15

Ordered pair 13
Ordered triple 13

Permutation 20
Product of cardinal numbers 48
Product of sets 14
Proper subset 40

Range of a mapping 17

Sequence 59
Subset 4
Sum of cardinal numbers 35
Surjection 18

Union 8